VITICULTURE FOR BEGINNERS

Basics of viticulture, planting a vine, viticulture in organic and biodynamic methods, grapevine diseases, pests that affect grapevines and many more.

BRANDON LUIS

PREFACE

These pages, which were written with love and shared knowledge, look into the details of taking care of vines, discovering the mysteries of terroir, and honoring the time-honored traditions that make viticulture a mix of art and science.

ACKNOWLEDGEMENT

Thanks a lot, to friends, family, and people in the grapevine community for always being there for me and sharing my love of the vineyard craft. May there become lots of clusters, good soil, and raised wine glasses forever in celebration.

Contents

CHAPTER ONE

BASICS OF VITICULTURE

Viticulture, which deals with the sequence of events that take place in the vineyard, is the cultivation and harvesting of grapes. Viticulturists use their knowledge of vine ripeness and quality to determine when to harvest grapes, control pests, fertilize, prune, and create sustainable farming practices.

Since humans first recognized and began to use grapevines, individual cultivars (varieties) have been chosen for a variety of qualities. These carefully chosen vines were refined, cloned, cross-bred, and hybridized over the many centuries that followed. As a result, at least 1,500 varieties of wine are currently known to be produced, with many more surely to follow. Since Phylloxera spread to Europe and restrictions were put in place on vine cultivation, grafting, and distribution, provenance and variety have become increasingly important. The variety or varieties from which a wine was made have taken on a much greater significance since the 1930s, when wine appellations were introduced. This was particularly true in the second half of the 20th century, when variety names began to appear on labels, giving consumers a simple, quick way to identify and interact with wine.

Grapevine clones are frequently further subdivided into clones, each of which has distinct (and sometimes very distinct) characteristics while still adhering to the general characteristics of the variety. The process of creating clones involves choosing individual vines with desired traits (such as greater disease resistance, increased yield, early ripening, deep color, small berries, etc.), cuttings from those vines, planting them, and then monitoring the vines to see if any of the cuttings have inherited the desired trait.

A new clone with an exaggerated version of the desired traits is created by picking the cuttings that do have those traits and repeating the process over a number of generations. While it may seem like a straightforward procedure, the process actually takes a long time, and there is never a guarantee on the result. A fresh clone will need to reach maturity before demonstrating its worth over multiple vintages.

Clonal trials are typically conducted by universities and research institutes. They test the plants at various locations and under various conditions, and only after thorough investigation are they authorized by the government and made available to commercial growers. Clones are occasionally chosen by specific growers, vineyards, or appellations. For example, many Italian growers treasure their local Sangiovese clones, and it seems that every village has its own clone of this widely grown varietal. For example, Brunello di Montalcino, a Sangiovese clone, is sometimes regarded as a different grape variety. There are a lot of individual clones in varieties that are grown in a variety of climates and locations.

Consider Pinot Noir.

This type is separated into clones that are suitable for warm and cool climates; clones that are suitable for different purposes, such as darker-skinned clones for producing red wine or lighter-skinned clones for producing (white) sparkling wine; and clones that are suitable for different wine qualities, such as lower-yielding clones for better sites or higher-yielding clones for basic red wine. Furthermore, there are clones with loose bunches (Mariafeld clones) and clones with an upright habit (Pinot Droit). Some growers in the New World prefer clones like the Mendoza clone, which has both large and small grapes and a distinct tropical fruit character.

Another variety, Chardonnay, has clones to produce wines with high acidity and relatively little fruitiness (such as might be used in Chablis). Clones are important enough that many appellations list both the variety of vines and the clones that can be used. This is particularly evident in the Champagne region, where eighteen clones of Pinot noir, eleven of Chardonnay, and eleven of Meunier may be grown, according to the CIVC (Comité Interprofessionnel du Vin de Champagne). New clones maintain the name of their parent variety, which is the one significant advantage they have over newly cross-bred varieties. Therefore, even though a new clone may differ noticeably from older clones in terms of flavor and quality attributes, from the standpoint of marketing and appellation, it is regarded as the same variety; hence, a new name on the bottle does not need to be explained to the (occasionally easily confused) public.

Bulk Clone selection

Not all growers agree with the idea of clones. Many people believe that a single clone vineyard lacks the complexity of a traditional multi-clone vineyard due to its uniformity in style and flavor, which could be viewed as simple or one-dimensional. Before phylloxera made grafted vine planting necessary, vineyards were rarely completely removed, or grubbed up. Alternatively, vines that died or reached the end of their productive lives could be replaced with layering.

Additionally, the grower would take cuttings and root them in place, possibly from the same vineyard or from a neighbor whose vines (of the same variety) had yielded good fruit. Thus, a Pre-Phylloxera vineyard might have included several, grower-selected clones from the best vines in that specific vineyard or area. Many old-world growers believe that the history of their appellation is captured in their multi-clone vineyards, which they believe contribute significantly to the terroir of their sites.

Establishment Of a Vineyard

A vineyard must be established with a significant financial investment, and once a vineyard is planted, it is typically not feasible (or financially practical) to alter its basic layout. The selection of varieties, clones, and rootstocks, planting densities, row widths, and trellising designs are among the mistakes that are most difficult to undo after the fact. Additionally, several pre-planting procedures, such as subsoiling, rock and stone removal, and drainage, can only be completed correctly and only before planting. Put another way, if you want the investment to be wise, you must get most of the fundamental details right the first time.

Precision viticulture (PV) is characterized by the use of spatial variation knowledge to inform decision-making. The invention of Global Positioning Satellites (GPS), which have transformed practically every aspect of life, including agriculture, is to blame for this. Utilizing GPS tracking devices, satellite imagery, electromagnetic sensors for field mapping, and earth radar, among other tools, farming is experiencing technological breakthroughs that were unthinkable just a few decades ago.

GPS is used to steer tractors and other powered equipment in straight lines. It also allows seed drills, fertilizer spreaders, sprayers, and harvesters to be programmed to react differently when their GPS systems detect their location in the field. In order to maximize yields, both horticultural and arable farmers now have digital maps of their land and harvesters that track yields in real time as they go across the fields. This technology enables them to plant at higher seed rates and apply more fertilizer to productive areas.

Similar to this, irrigation systems can be adjusted to supply water to those parts of the land that can respond to higher watering rates, produce larger yields, and ensure that valuable water is not wasted. Drones are also beginning to appear in agriculture, including viticulture, for a number of purposes; the most popular ones in vines are vigor mapping and disease monitoring. While still uncommon in viticulture, PV is increasingly used and will eventually become much more common. Monitoring the spread of Phylloxera in the Napa Valley was one of the earliest applications of infrared imaging, which was first achieved by taking images from a light aircraft.

A vineyard's Normalized Difference Vegetation Index (NDVI) provides a highly accurate picture of the vines' vigor. Higher vigor areas appear green, while low vigor patches of vines (those most likely to be damaged by the insect) appear red. The suspicions can be verified by on-site inspections, and the pest's development can be tracked over a few years. The Leaf Area Index (LAI) of the vine, which is the ratio of leaf area to soil surface area, can also be well-represented by the NDVI. The higher the figure, the greater the leaf area and, consequently, the greater the vigor of the vine. Growers can map their fields ahead of any preparation work and divide them into areas of soil type (usually into areas where vine vigour is the determining factor) that may require different physical treatments: fertiliser requirements, irrigation systems, rootstocks, varieties, and clones. Site surveys using electromagnetic sensing and earth (or ground) radar, coupled with physical soil sampling and testing, enable growers to map their fields. Better, more economical crops will be harvested if the proper vines are planted for the soil in which they are growing. The foundation of PV in vineyards is the mapping of vine growth stages and vigor. From there, all treatments applied to the growing vines, such as pruning loads, canopy density, canopy management techniques, pest and disease control, irrigation, and harvesting timing, can be adjusted to maximize crop quality and quantity.

Large-scale contemporary vineyards will eventually employ these methods more frequently as an increasing number of sites are surveyed, planted, and then monitored. In top and soft fruit orchards, plantations, and vineyards, GPS-controlled self-driving sprayers—or tractors pulling sprayers—are becoming more common. Once trained a route for spraying, the sprayers will automatically follow it spray-round after spray-round, allowing one operator to operate two sprayers. Orchards and vineyards are also introducing self-driving mowers. These days,

GPS vine planting equipment is commonplace, resulting in incredibly precise row widths and intervine distances.

Site preparation: Before planting, many sites require some regrading. This could involve doing minor adjustments like filling in a few strange depressions and removing a few humps and bumps completely reforming a north-facing slope to a south-facing slope by removing all of the topsoil, followed by a significant amount of subsoil removal (which was in this case used to cover their new 100,000 barrel barrel hall with three meters of insulating soil), regrading the slope, and adding topsoil back to the site. But extreme gardening like this is not common. The process known as Flurbereinigung, or "parcel cleaning," has transformed many of the old river-facing terraces on the Rhine and Mosel (and the other vineyard rivers in Germany), dramatically changing both the appearance and the viability of grape growing in these areas economically (it was later described to me as "the way the Good Lord would have wanted it had he had the money"). Before this reorganization, a typical grower might own numerous relatively small parcels of vines spread out over several kilometers. These parcels were typically only a few rows long, with as few as ten vines per row, and they were all challenging, if not impossible, to mechanize. Each parcel was served by a narrow walkway that all labor and supplies would need to travel up and down, respectively, and that all grapes would need to be hauled down. To put it briefly, these vineyards required a lot of labor and were not profitable, unless high wine prices were involved. Gathering growers, the Flurbereinigung commission devises a plan to remove the old terraces, erects tall walls and wide roads, installs a main water supply for water spraying, and generally realigns and regrades the land. Then, growers are returned to their vineyards in larger individual parcels distributed as evenly as possible over the entire area (though a smaller area overall as some land is inevitably lost to the new roads and turning areas). As a result, growers have access to drive-to vineyards, mechanized vineyards that can be utilized with winches mounted on the sides of tractors or crawler tractors, and vineyards from which grapes can be directly loaded onto a transport vehicle and transported back to the wineries. Due to this, farming at these special riverbank locations has become much more cost-effective, and wine production has been maintained.

It will be necessary to subsoil and drain a lot of sites before planting. The subsoiling tines must be adjusted so as to avoid disturbing the drains since subsoiling is typically done after draining. Nonetheless, this may very well happen before draining in places where rocky or compacted soil regions are known to exist. In this case, before more work can be done, any exposed rocks or very large stones will need to be removed from the site. If the site is going to be irrigated, now might be the best time to install header pipes and mains, the foundation of a water supply, to prevent soil disturbance later on.

In the event that removing older vines is necessary before planting new ones, it's critical to remove as many of the older roots as possible by burning them or removing them from the area, and to thoroughly rip the area to disturb any remaining roots. A period of fallowing which involves planting a green manure crop that can be ploughed back into the soil, may be necessary for these sites. This will help clean up the area and replenish the soil with humus and fertility. A season of fallowing, in which weeds can germinate and be sprayed off with weedkiller, followed by a crop of green manure, is often a good idea and a way to get a cleaner and healthier site before vines are planted, even with sites that have not previously been planted with vines and where time allows. Before planting, sites containing harmful perennial weeds can be treated with hormone weedkillers (e.g., 2,4-D), the use of which in established vines would be unethical, illegal, or both. It is important to take precautions to ensure that these highly potent weedkillers don't drift onto nearby vineyards and cause harm. Farmers who practice organic and biodynamic farming, and are prohibited from using weedkillers, may need to dedicate two growing seasons to sowing cover crops and clearing their land in preparation for planting. Following completion of the aforementioned procedures, sites must be tilled to create a smooth tilth at permits planting. In order to be closer to the rooting zone after plowing, which turns over and buries the surface layer, farmyard manure, compost, and fertilizers are frequently applied before the site is ploughed.

Ploughpan

Over several years of plowing land at the same depth, an impervious layer of compacted soil may have developed in some locations. In order to remove any hard compacted soil—or even subsurface rock—where this happens, crawler tractors equipped with multiple deep ripping tines—typically ranging in length from 500 to 900 mm—mounted on the back will be utilized to tear through the soil from one side of the site to the other and then again at right angles. After this rather grueling procedure, any large rock fragments may need to be removed before plowing at a more conventional depth (200–300 mm), at which point regular cultivation can resume.

Drainage: As was previously mentioned, artificial drainage may be required in some soil types, in areas with abundant spring and summer rainfall, or in situations where vines are to be irrigated. Even after vines have taken root, a site may be drained using a mole drain—a torpedo-shaped device that, when dragged through the soil, creates a self-supporting drain—in soils with a high clay content, provided the site's slope and planting orientation permit it. Mole drains may last for several years, but eventually they will need to be renewed, depending on the type of soil.

However, before planting, a network of perforated plastic pipes will be installed below soil level in the majority of soils that need drainage. (Prior to the 1980s, porous clay pipes, also known as "tile drains," were used.) A number of variables will affect the drainage system's actual design, including the soil type and annual rainfall, which will determine the depth of the pipes and their spacing. However, pipes are normally installed with a distance of 10 meters between drain lines and a depth of 0.80 to 1.00 meters. To make sure that all pipes are positioned to fall from high to low and that any drainage water drains away into an appropriate watercourse, a drainage machine guided by lasers will be utilized.

Drainage is typically only possible prior to planting and is not a cheap option. Draining a site before planting is therefore a better idea than regretting it later. The superior natural drainage of the top estates on the left bank of the Gironde contributed to their long-term success over those on the right bank. But after the Bordeaux Classification of 1855 (which only included properties on the left bank), the fortunes of the leading châteaux turned around, and their increased wine revenues made it possible for them to construct drainage systems in their vineyards. Top right bank properties (Petrus, Lafleur, Cheval Blanc, Ausone, etc.) have only been able to drain their vineyards to improve their growing conditions and, consequently, their fortunes, in the last few decades (actually, since the end of the 1939–1945 war). Green manures, also known as cover crops, are crops that are intentionally planted to suppress weeds and provide organic matter that can be tilled into the soil to increase its organic matter content and eventually improve its humus status. For a multitude of reasons, the precise kind of cover crop that is planted will vary; however, it may consist of mixed cereals like wheat, barley, oats, and rye; legumes like mustard, peas, beans, vetches, and clovers; or simply fast-growing grasses.

Other possible varieties include root vegetables like turnips or fodder radishes, which have deep roots and aid in soil disturbance. They will typically be allowed to reach their full height before being ploughed in if they are sown well before planting. (This method can also be applied to already-existing vineyards; however, caution must be exercised to avoid starving the vine of water and nutrients.) Certain plants, like oilseed radishes, mustard, and rapeseed, can function as biofumigants, thereby decreasing the quantity of nematodes that carry viruses.

Windbreaks: A lot of vineyards are exposed to the wind, which cools the vines, slows down their growth, and delays ripening. The vine is a sugar-producing plant that produces less sugar when its leaves are moving. Better photosynthesis and, thus, better ripening are correlated with stiller leaves. Creating a natural windbreak, rather than a man-made one, could be one of the pre-planting tasks. The windbreaks will be large enough to offer some cover by the time the vines begin to crop, and they will get better as they mature. In order for the windbreaks to have some growth and be offering a decent degree of protection by the time the vines begin to crop, windbreaks are occasionally planted a year or two ahead of the vines. The climate and location of the trees will determine which kinds are best for windbreaks.

The fundamental guideline is that the trees have to be in leaf before the vines flower and continue to do so until harvest time. They should also be compact and upright, able to be lopped once they reach the proper height, and have a deep and robust root system that will keep them from being blown over in strong winds once they reach maturity. They should also not have an overly pervasive root system that could deprive nearby vines of water and/or nourishment. Impervious evergreen conifers with year-round foliage are typically not satisfactory as an effective windbreak should filter the wind rather than stop it.

Additionally, birds and their nests are more likely to be found on evergreens than on deciduous trees. As a general rule of thumb, a windbreak will shield an area ten times larger than its height; for example, a vineyard that is 100 meters wide will be protected by a windbreak of suitable trees that is 10 meters high. In dry regions, living windbreaks might need drip irrigation until they get established. They also typically need to be protected from rabbits, hares, deer, and other animals, and they need to be given the same care and attention as the vines they are meant to shield from weeds. Once established, a carefully maintained windbreak can be a helpful tool in enhancing the vines' microclimate.

Synthetic netting and various plastic strips can be erected along the windward side of the vineyard as alternatives to living windbreaks. However, in order for these to withstand a winter gale, a robust system of posts, wires, and anchors is needed. Larger vineyards may need internal windbreaks to divide the vines into distinct blocks because they cannot be protected by a single windbreak around the windward perimeter. This can be an extremely successful method of enhancing the microclimate, provided that the windbreaks are not overly tall and do not shade the vines to one side of them.

However, natural hedges are typically the simplest—as well as the most visually pleasing—option. In an attempt to protect the vines and keep birds out, some vineyards have experimented with completely netting the vineyard, including the overhead and side walls. It has been demonstrated that installing all-over netting significantly raises the heat summation in wind-exposed vineyards, producing riper grapes and better wines. The netting that is used overhead needs to be selected so that it doesn't obstruct too much light and has a mesh big enough to let snow (if it falls) fall through it instead of collecting on top, which would collapse the entire structure.

Hail netting: While it is comparatively uncommon on vines (it is much more commonly used on soft fruit, stone fruit, and citrus), hail netting is present in areas where this issue is predicted. Mendoza in Argentina and North Italy are two examples. In order to avoid obstructing the yearly vineyard work, netting is installed on a system of posts and wires that are (usually) independent of the posts and wires of the vine. The netting is virtually always left up all year round due to the enormous cost and time required to put it up and take it down, as well as the fact that hail, while more frequent in the middle of summer, does not always follow a convenient schedule. A cluster of grapes can easily be damaged by a hailstone as small as 10 mm. Netting must be strong enough to withstand the shock of a potentially powerful hailstorm, strong enough to stop even small hailstones, and light enough to allow the vine to grow unhindered. A slight reduction of 5% might be too much in colder climates; 15% might be appropriate in warm/hot climates. Typically, the netting is set up in a "plough and furrow" pattern, or a sequence of Vs, with a small space between each join at the bottom of the V in the middle of the row between the vines. As a result, the hail strikes the inverted

V's peak, which is directly above the vine row, and then it slides through the opening and onto the vineyard floor by sliding down the sides into the furrow, which is directly over the alleyway's center line between the vines. This removes the hail and prevents it from falling onto the net, which would surely shatter the net, put undue strain on the support system, or both. Heavy snowfall can also damage the netting and its supports if the net's weave is too tight and the area is susceptible to snowfall. Installing hail netting has additional benefits. It can offer some shade in extremely hot climates; in windy climates, it will give the vines great growing conditions and raise their ripeness levels; and in all climates, it will keep birds away, though extra side and end nets will typically be required in the final 6–8 weeks of the growing season.

Protection from predators: For a number of reasons, perimeter netting is frequently either impractical or unfeasible. Many growers believe that it's worth it to lose a few vines to predators because maintaining the netting takes too much time. However, where it is practical to do so and where land-based predators such as rabbits, hares, rodents, badgers, deer, wild boar, and others are known to pose a threat, it is frequently more economical—and certainly more successful—to enclose the entire site with a net. A minimum of 900 mm of netting must be erected above ground, and an additional 300 mm must be buried in the ground, with 150 mm positioned vertically and 150 mm tilted outward from the vineyard, to deter rabbits and hares. This will stop bunnies from making tunnel entrances. To keep out the kind of deer present in the area, deer netting must be at an appropriate height, and a wire-mesh fence with a height of 1.75 to 2.40 meters is typically needed. The only option may be to use netting and local culling (where permitted) to protect the sites from more powerful predators, such as wild boar and badgers, who see fences as challenges rather than barriers. When using perimeter netting and the plan calls for using herbicides to control weeds, each vine should ideally also be shielded by a 500 mm high spray-guard until its trunk is strong enough to withstand an herbicide spray, which usually happens after two years of growth. If you are not using perimeter netting to keep small pests like rabbits and hares away, you will typically need to have a guard for each vine. These are typically composed of plastic of some kind and are solid-sided (some have ventilation holes in them) or netted (of a synthetic material or metal chicken-wire mesh). There are numerous variations: some are spherical, some square or triangular, some have an opening side panel, and some just slide over the top of the vines.

Wax-coated cardboard sleeves (which frequently resemble milk or juice cartons with incorrect printing) are quite common in certain countries. Regardless of the kind, they all function as miniature greenhouses that, depending on the climate, either encourage the growth of the young vines or occasionally fry them. The presence of ventilation holes in the guards helps prevent fungal problems, but the guards themselves can become extremely humid. The majority of guards are very difficult to use, need to be tied to a stake or cane, and increase the difficulty and expense of side-shooting and shoot selection. Although regular maintenance and patrolling are necessary for the perimeter netting solution to remain effective over time, that is always my first choice. If growers choose to use individual plastic guards, they will benefit from protection from weedkillers during the first two years of vine establishment (roughly by year 3). After that, the guards can be removed and, if UV exposure hasn't caused them to disintegrate, repurposed in another new vineyard.

Frost Detection

Winter frost: There isn't much practical protection against extreme winter frosts, which are defined as frosts that fall between December and February in the northern hemisphere during midwinter and cause temperatures to drop to between -20°C and -25°C. Grafts can be earthed up for protection (buttage), as previously explained, but this provides no defense against the vine's trunk and canes. Extremely significant damage will occur when the vine experiences these kinds of temperatures and its woody sections have some moisture in them, which naturally freezes.

In extreme years, like 1956 in Bordeaux, winter frost completely destroyed large areas, and for many properties the only way to recover was to replant. Even extended periods of mild winter frost can exacerbate conditions like Eutypa lata42 and Crown Gall, which are caused by the bacterium Agrobacterium tumefaciens. Unprotected vines will frequently die when temperatures drop below −25°C. As was previously mentioned, in order to protect vines from extremely cold winter temperatures, in certain parts of the world, vines are placed on the ground and covered with soil or a protective cloth before snowfall. Spring frost: There are two types of spring frosts: advection and radiation frosts. These frosts happen in April, May, and sometimes even June (in the northern hemisphere) when temperatures drop to -1°C to -6°C. When extremely cold, frost-carrying air blows into a vineyard area—sometimes from a great distance away—warmer air is displaced, resulting in an advection frost. This kind of frost is more typical of winter weather and is uncommon to see in the spring. Steer clear of frost pockets as the only way to fend off an advection frost.

The second kind, called a radiation frost (also called an inversion frost), usually happens on a clear, calm, dry night when there are no clouds or significant amounts of water vapor in the air to obstruct long-wave radiation from the ground to the air, and the air around the vines cools down quickly. The term "inversion" refers to the situation where a layer of cold air is trapped beneath a layer of warmer air and is unable to escape. This chilly, harmful layer usually begins nearly at ground level and rises to a height of 15 to 30 meters above the vineyard floor. This is the most common kind of frost that damages vines and is also the easiest, albeit priciest, to manage. Choosing a location away from a likely-to-frost area or, if frost is a possibility, planting the vineyard so that cold air can drain downhill towards an unplanted area where natural ventilation will allow the frost to be blown away from the vines are the best defenses against any kind of spring frost. If this isn't feasible, there are a few safeguards that can be implemented. Frost protection techniques include a variety of heating techniques for the vineyard, such as using burners that run on gas or oil, or even individual 5-liter paraffin wax candles, known as bougies in France. Smudge pots, or burners that burn a mixture of waste sump oil and paraffin, were once common but are now uncommon due to air pollution concerns. Furthermore, reports frequently surface about growers burning bales of straw and hay soaked in used oil, or even lighting the tires of old cars and trucks in a desperate situation. Any kind of burner should be positioned thoughtfully throughout the vineyard and lit gradually as the temperature drops. In addition, there are electric heating cables strung along each row, tractor-powered fans that gather cold air and blast it upward into the warm air layer to create air circulation, and tractor-mounted air blast fans that blast hot air heated by gas or oil into the vineyard.

Helicopters are used in New Zealand to hover over the vines and accomplish the same purpose as windmills. Other methods include the use of massive windmills that mix the hot and cold air layers. Depending on the severity of the issue, a vineyard owner will frequently use two or more of the aforementioned methods—oil burners plus a frost-mill. A different method that is frequently applied to citrus and other soft fruits is misting the vines with fine water mist as soon as the temperature drops below freezing. After that, this water freezes on the buds and canes, shielding them from harm. As unlikely as this method may seem, it works as long as there is enough water available. The little amount of heat that is released by freezing water—referred to as the latent heat of freezing in scientific circles—along with the ice layers that form around the canes and buds create an igloo effect that shields them from harm. The water sprays can be turned off once the temperature rises above freezing, and as the day warms, the ice gradually melts. This method, also called the aspersion technique, uses enormous amounts of water—up to 25 to 30 mm per night. If stored water is used instead of water from a river or canal, it's critical to have enough water on hand to maintain the sprays operating for the maximum number of days and the maximum depth of frost that is feasible. Running out of water in the final few hours of a six-day frost period is not a good idea. Additionally, it is crucial to prevent the soil from becoming soggy; soils with inadequate drainage may suffer in this regard. This method is unpopular in areas where water is limited and will be needed in the summer because of how much water it uses.

All of the aforementioned tactics, which entail immediate human and material expenditure during the frost, could be referred to as active methods of frost protection. While remote monitoring systems that send texts and emails can at least alert one of impending frosts, constant monitoring during likely frost periods is required. Individual vineyards, which may well be some distance from habitation, will have to be visited and temperature readings taken throughout the night when a frost is anticipated. Nonetheless, there are a number of inactive steps that can be done to lessen frost damage—possibly without totally eliminating it. When it comes to the risk of frost, the condition of the vineyard alleyways matters. Unlike a vineyard covered in grass or other vegetation, clean, clear, bare earth retains heat better during the day, especially if it is slightly damp. During the evening and night, this heat is then released into the atmosphere, assisting in the prevention of frost.

The worst condition and one that is most likely to worsen frost damage is an unmown, long grass sward that retains moisture. Many growers whose vineyards are vulnerable to frost use contact weedkiller in the early spring to burn off all vegetation on the vineyard floor, allowing it to regrow after the frost-hazardous weeks pass. It will be beneficial to train vines to a high-wire system like GDC, Sylvoz, or Blondin43 because the soil is always the temperature that is lowest there. At 1.50 m above sea level, the temperature can be up to 1°C higher than at 0.75 m. Frost-prone areas can benefit from late spring pruning because pruned vines will have a delayed bud-burst and possibly less frost damage. Some vine growers will leave extra canes, or "sacrifice canes," on their vines. If frost damage has occurred, they will leave the extra canes to carry a crop; if not, they will remove them. Planting cultivars like Meunier, which naturally emerge from dormancy later, may also be beneficial. Growing some varieties (like Pinot Noir and Chardonnay) in frost-prone areas may be worthwhile because they exhibit a notable tendency to produce viable secondary flowers after the primary ones have frozen.

Additional tools include sprays that protect vines from light frost, such as those based on liquid acrylic polymer (Anti-Stress, Frost Protect, Frost Shield, and widely used in citrus plantations). Enormous nets and covers above (possibly for use against birds and/or hail) will provide significant frost protection. In certain areas, insurance coverage against frost can be obtained similarly to that against hail, although this is typically a costly choice. Vineyard layout: There are a number of factors to consider when laying out a vineyard, and no two sites are the same when it comes to the direction, length, and location of the rows and access roads. Nonetheless, there is typically one standout feature (e.g., the site has one strategically placed boundary that makes a certain row orientation obvious or the site is surrounded by neighboring vineyards whose rows run in one direction) that simplifies the decision. Planting rows in a way that runs up and down a slope is generally a good idea, especially if the slope is at an angle higher than roughly 10 degrees, as this will prevent tractors and other machinery from sliding sideways. Unless a more easterly or westerly-facing vineyard is preferred for local reasons, rows should ideally run north-south. In order to receive more afternoon sun in the northern hemisphere, some growers choose to plant their rows more south-westerly. The idea is that the vineyard will have warmed up by then and will continue to remain warm well into the early evening. Some people favor orienting their rows towards the southeast, believing that this direction will cause the rows to warm up faster and evaporate any overnight dew. In other circumstances, rows facing 90 degrees to the predominant wind may offer more favorable growing conditions since the first few rows of vines in the area block cooling winds. These row orientations are the opposite of what happens to vineyards in the southern hemisphere. The best alignment for the rows, from a purely practical and economic standpoint, is parallel to the longest boundary of the site or to an existing feature, like a road, hedge, or neighboring property. This will guarantee that the rows are as long as possible (within reason) and that there are as few rows ends as possible—each of which needs a straining post and anchor. Additionally, fewer short rows will need to be mowed or sprayed, reducing the number of turns needed. Nearly all vineyard layouts are a compromise between ideal conditions for the vines and pragmatic factors like the site's natural slope or existing boundary. Even though the rows may face almost west, it is preferable to plant straight up and down a slope rather than planting them facing due south, which could result in issues with machinery slipping across the slope for the duration of the vineyard.

CHAPTER TWO

PLANTING A VINE

Planting should be timed according to the site and local climate. It is possible to plant vines from early winter to late spring in warm climates where frosts in the winter and spring are not a concern. Until they are distributed, nurserymen typically store the vines in cold stores; by the time the vines reach the vineyard, they should be dormant. Although they are best planted as soon as they arrive, they can be stored unplanted in cool, damp conditions for a week or two until needed for planting. It is best to wait to plant until the risk of spring frosts has passed if this is likely to be an issue. When planted in March or April of the same year, vines planted as late as May or even June (in the northern hemisphere) will typically catch up with each other, assuming the soil is sufficiently moist. The ease of planting vines is often underestimated, as the amount of work involved in maintaining them after planting is done is substantial. Although many supplying nurseries will do this for the grower, most vines need to be prepared before they are planted. Every vine should have its upper growth reduced to a single bud. For certain planting systems described below, the roots may also need to be heavily trimmed to prevent folding back when the plants are inserted into the ground.

Typically, the nursery rewaxes the upper quarter of the vine with soft grafting wax to keep moisture in, shield the vine from fungal invasion, and prevent it from drying out. A planting system that permits the roots to be left at approximately 100 mm length is preferred by most growers because trimming the roots very short (shorter than 50 mm) always seems to make the vines slower to get off the mark after planting.

In warm climates, it may be necessary to water the vines after planting; otherwise, they should be kept damp and out of direct sunlight until they are ready to be planted. Prior to planting, the planting hole may occasionally be filled with peat and fertilizer. In order to encourage the vine's roots to interact with the soil as soon as possible, planting dips containing mycorrhizal fungi is also becoming more popular. As the vine is planted, some planting machines will pump fertilizer and/or water into the planting hole. In hand-planted vineyards, precise and organized planting will necessitate marking out each vine's position before planting. At times, markers are only used to mark the ends of the rows; the distance between each vine is then measured by stretching a wire that has markers at the location where the vine is to be planted between the end posts. Some growers just mark the positions of each vine with a planting cane and a tape measure. Regardless of the chosen method, it is more effective, simplifies life, and appears more organized when vines adhere to a precise grid pattern. The actual process of planting the vines can be done in several ways: a large dibber44 can be used to create a hole specifically for each vine; a spade can be used to dig a good-sized hole for well-rooted vines; and, in the right soil conditions, a water-lance that digs holes under water pressure is a very quick method.

Additionally, vines can be machine planted using a laser-guided or GPS-guided planting machine, which can plant at very precise row widths and intervine distances and requires little to no pre-plant marking out, which saves a significant amount of money and time. Regardless of the system employed—which will actually depend on the size of the operation, the site's condition, and the quality of the soil—each vine should ideally be firmly planted, with as little space between the roots as possible and with the roots nicely spread out and not pointing upward. Although this may not always be feasible with machine-planted vines (a lot will depend on how the site has been prepared as well as the type and moisture state of the soil), the benefits of quick, economical, and precise planting without the need to mark the site out more than make up for a planting machine's heaviness. Given adequate water and care during their initial years, vines are fairly resilient plants that, unless mistreated before planting (such as being allowed to dry out, exposed to excessive light, or exposed to extremely high temperatures), almost always survive and mature into mature plants. It is important to plant grafted vines at the proper depth, with the ideal graft being approximately 50 mm above the soil's surface. When planting a vine with the graft too close to the ground, there is a risk that soil will accumulate against the graft (vines frequently settle deeper in their planting holes after being planted) and that unintended, extremely vigorous shoots from the rootstock variety and/or roots from the scion variety may grow at the graft point. Naturally, these roots, which originate from the vinifera scion, are susceptible to Phylloxera. The method of controlling weeds must also be considered when determining the actual height of the graft above the soil. For example, growers who use mechanical weed hoes prefer their vines to be slightly higher than those who use herbicides. This can result in scion roots, as previously mentioned, since many mechanical hoes leave a mound (or hill) of dirt surrounding the vine. After planting, care for vines: After planting, vines go dormant for a week or two, depending on the temperature at which point they begin to bud. Little green shoots will emerge as the buds begin to push. They are now ready to begin their training. Young vines can be trained using a variety of methods, and different growers will have different opinions about the most effective ones. Dry-climate growers who don't mind their vines taking their time to establish (and who don't want to spend a lot of money on trelliswork the first year) will frequently leave their vines almost entirely alone for the summer following planting, allowing them to produce several shoots that can be clipped back in the winter. All they would do for the vine would be to keep weeds out of its way, cultivate the rows, tie up new shoots that might fall to the ground, and, if needed, shield it from pests and illnesses.

There are other growers who will believe that the vines need to be trained as soon as possible—up something. In this situation, stakes made of wood or metal, bamboo canes, or strings fastened from the vine to a wire are possible solutions. Regardless of the kind of support structure employed, the goal remains the same: to initiate the process of developing a solitary, robust shoot that will eventually develop into the trunk and give rise to the first fruiting cane or canes. To begin with, the young vine will frequently produce a number of small shoots, all of which need to be removed by rubbing or cutting off, but one. The ideal shoot to be chosen is the one that will grow straight up from the rootstock and eventually become the trunk; this shoot is frequently not the strongest grower. After being chosen (all other buds and shoots having been rubbed off), the single shoot is supported in growing straight and true by the removal of its side shoots, which directs the vine's entire energy toward it. Till autumn, when the new vine may have reached a significant height, this work will be done. In vineyards that are not experiencing a water deficit, 1.5–2 m of new growth is typical. As the cane grows, it must be tied to the chosen stake in order for it to mature. Simple metal twist ties covered in paper are one type of tie; more complex clipping and fastening tools are available to expedite the process. Regardless of the kind of tie used, it should be able to swell and expand as the cane becomes a stem and then a trunk. If individual rabbit guards have been used, they must be raised and then replaced (or undone, depending on the guard's specific design). One reason not to use this kind of rabbit protection instead of the whole-vineyard perimeter netting method is this additional work. The strain will be too great at the end of the first growing season, before the vine sheds its leaves, and especially if the leaves are wet and there is a strong wind. If the vine and the stake are left unsupported, they will probably fall to the ground. It is best to install the entire trellis system in the first few weeks after the vines are planted; in that case, the individual vine stakes can be permanently tied to the first wire and integrated into the trellis as a whole. For as long as the vine requires it, the first-season training should go on.

Weed Control in Vines

Managing weeds is unquestionably one of the most crucial aspects of caring for recently planted vines. In addition to physically crowding out and occasionally suffocating the vine, weeds will deplete the young vine's supply of water and nutrients. Due to the damp weed foliage impacting the fragile vine, this frequently results in an increase in disease problems. Many times (though not always), the precise technique employed to maintain weed-free, freshly planted vineyards will be similar to that of the grower's established vineyards. There are mechanical methods as well, like hand hoeing or strimming; using a tractor-mounted hoe or rotary cultivator; or even using a gas-powered flame thrower to burn off. Mulches—both natural and synthetic—can also be used. Another popular and highly successful option is herbicides.

Alternatively, you can leave a permanently mowed grass strip beneath the vine; however, this will usually only be done once the vines are well-established and have access to water. Before a weed control system is implemented, the two stages really need to be taken into consideration together, as growers have access to all of the aforementioned options once the vineyard is established and begins to produce. It takes a lot of time and money to hand-hoe, and a poorly-aimed hoe can quickly harm a young vine. Even though it might not cause the vine any immediate problems, a wound to the side of the young trunk frequently results in a lesion that could potentially prove fatal in the future if something like Crown Gall gets into the wound. Trunk diseases may also become more common as a result of hoeing and strimming wounds. Since young grafts are extremely delicate, a hard hit with a hoe will frequently destroy the vine. Strimming is a time-consuming method that is typically reserved for situations in which there are few vines that need to be removed; it is also prone to harm the delicate first year's growth. Additionally, with both of these techniques, the vine will have lost some of its water and/or nutrients by the time they are applied, which occurs after the weeds have spread. Using a gas-powered flame thrower to burn weeds is a seldom used method that seems to be more common among organic growers. However, it is not suitable for use on young vines that have delicate stems. In well-established vines, it works well, but it must be done often to prevent the entire vineyard from catching fire!

Numerous varieties of tractor-mounted under-vine cultivators and mowers function by means of a revolving blade cultivator or mower that swings out and away from the vine in response to a sensor that triggers a hydraulic or spring-loaded arm that the implement is attached to. In order to prevent the vines from rocking and becoming loose in the ground over time, this weed control technique requires that each vine be firmly fastened to a fairly sturdy support. Furthermore, the sensor on the swing arm has a tendency to strike the vine as frequently as it strikes the support stake, which can cause lesions and bruises that could develop into issues down the road. Climate is the most obvious of several factors that will determine how often a tractor needs to go through the vineyard in order to maintain it free of weeds using a mechanical hoe. It may be necessary to go through the vineyard six or eight times during a season in areas that receive summer rainfall, as weeds may not immediately wilt and die when uprooted. In warmer regions, this could be lowered to three or four passes. A tractor that has a rear-mounted mower or cultivator and a mid-mounted hoe will be able to do at least two tasks simultaneously.

Mulches have their benefits, whether they are made of natural materials like chopped straw, peat, or chipped tree bark; inert materials like stone or crushed shells; or synthetic materials like plastic woven material or polyethylene film laid in a continuous strip. While loose natural mulches may cause nutrient imbalances that need to be corrected, they will eventually decompose and raise the humus levels of the soil. They won't provide 100% coverage, though, and a lot of weeds will still grow through and show their heads above the mulch. This means that extra mulching to suffocate them—or labor-intensive weed removal—will frequently be required throughout the growing season. They can be aesthetically pleasing and visually pleasing, though, and they will retain moisture naturally. In certain hot climates, a method called "mow and throw" (also called "slash and throw") is used. This method is more commonly applied to older cropping vines than to younger ones, and it involves cutting any vegetation growing in the middle of the rows—such as weeds and grasses—with a mower before throwing the material beneath the vines. Continuous plastic or fabric mulches can be applied before the vines are planted. The vines can be planted through the mulch using a water lance or dibber type of planting system, with the edges buried in the ground and the remaining 400 mm visible. It can be difficult to maintain the space between the buried edges and whatever is being done in the middle of the rows—cultivating or mowing, for example—because some mulches tear easily. However, this kind of mulch offers nearly perfect weed control and retains the majority of the available moisture. These kinds of mulch are also inappropriate for high-stone soils because the stones can tear the mulch and cause damage from rodents and birds that gnaw through it in search of food. It's also a rather pricey option.

The type of material chosen will ultimately begin to decompose and fragment. If it isn't torn up and removed from the area, it will begin to blow around and gather in unsightly heaps in field corners and up against fences and hedges. Even though it takes a lot of time and money to remove, it is at least a one-time task. It will then be necessary to use another kind of weed control system after removal. The last option, chemical weed control, is applied to a wide variety of crops and has advantages as well as disadvantages. Economy and ease of use are the advantages. A single tank-mix spray that contains a residual weedkiller, which stays in the soil and prevents weed seedlings from germinating, and a contact or systemic weedkiller, which either burns off (the contact type) or kills off by penetrating the plant's root system (the systemic type), combined with the right product(s) for the anticipated weed spectrum, applied at the right time, will keep the soil free of weeds throughout the growing season—at least as clean as it needs to be, maybe not completely free of weed growth. One or two weeks after flowering is usually a good time for a second spray in midsummer, but this is usually necessary in rich soils and in cooler (and therefore moister) regions, especially where there may have been previous crops where weed control was not a high priority (pasture, for example). In order to keep the soil clean during the dormant season, a winter clean-up spray is frequently used to burn off any weeds that have escaped. Compared to mechanical methods, weedkillers are much more economical because they require less tractor and man-hours and can be applied quickly. Chemicals are used widely and all over the world because, despite the materials' apparent high cost in comparison to other systems, they are actually a more affordable option. With each vine protected by an impermeable plastic planting tube or sleeve, growers can use chemical weedkillers during the first two years of growth with much less concern for damage. One drawback of using chemicals to control weeds is that they can easily harm or kill the vine if applied improperly, especially in the first year or so. In the first year of planting, growers often employ a very light approach (perhaps applying only a residual and a mild contact weedkiller, both at half-strength) along with some manual weeding and trimming before implementing a full program of weedkillers in the second and subsequent years. Residual weedkillers may work less effectively in some soils, particularly those with low levels of organic matter. The prolonged application of chemical weedkillers may eventually cause the soil's humus levels to decline, though this can be remedied by sporadically mulching the soil. Herbicides are not always thought of as the most environmentally friendly products, which is another drawback. Consequently, non-chemical weed control techniques are occasionally preferable due to their perceived as "sustainable."

CHAPTER THREE

VITICULTURE IN ORGANIC AND BIODYNAMIC METHODS

In order to protect against pests, diseases, and other viticultural issues, organic and biodynamic grape-growing methods rely less—and in some cases virtually never—on chemical intervention. Instead, they place more faith in the vine's capacity to self-sufficiency when given the proper support. An intriguing question that continues to be important to those who doubt the philosophy underlying this area of agriculture is whether this ability is intrinsic and found in all vines, or only in those that receive the stimulation of organic and/or Biodynamic practices. A certification body that is officially recognized (usually by the national ministry of agriculture) in their country must inspect all organic and biodynamic growers, or at least all those who want to label their produce as such. Before the land can be certified as organic, an inspection will be conducted to ensure that it has been free of weedkillers and other prohibited substances for a period of three years. Every country has multiple certification bodies, and while their standards may vary, some seem to be less strict than others, they all share two fundamental guidelines for certification: no genetically modified organisms (or very little above a very small threshold) and no chemically synthesized weedkillers, insecticides, fungicides, or pesticides. Though not entirely devoid of what most laypeople would consider "chemicals," many organic and biodynamic growers will use naturally occurring materials like sulfur and copper (subject to strict volume limits; see below). The Demeter organization, a global organization, oversees growers who practice biodynamic farming, which adheres to stricter standards than even organic farming. For example, crops grown beneath power lines require "extra applications of the full complement of the Biodynamic preparations" to counteract the EMF (elector magnetic forces) originating from the lines because "Biodynamic agriculture works with the dynamics of subtle forces."

Organic and Biodynamic practices: Prior to August 1, 2012, "wine made from organic [or Biodynamic] grapes" was the only type of wine recognized as organic or Biodynamic in the EU. Regulating 203/2012, on the other hand, went into effect on that date and allowed winemakers who followed the guidelines to label their wines as organic or biodynamic. A wide range of topics were covered by the regulations, which included the type of tank (which could not be made of plastic or fiberglass), the kind of yeast (which could not be genetically modified and had to be natural in the case of Biodynamic wines), the sources of materials used in the winemaking process (which had to be organic or Biodynamic), and much more.

Both organic and biodynamic farmers adhere to strict guidelines for managing their soils. They also use composts and manures that revitalize their soils, and in the case of biodynamic farmers, they also use a variety of "preparations" that improve the manures and composts they use and support the more organic functioning of the vineyard ecosystem. To put these "preps" in a human context, they work similarly to probiotics, which aid in the growth and restoration of good bacteria in the human digestive system. Biodynamic farmers assert that their preparations, which serve as compost starters, produce composts with a higher humus content than conventionally produced composts. Certain preparations bear similarities to human homeopathic remedies, as they depend on water's ability to retain a "memory" that conveys a signal to the plant.

As another method of protecting vineyards, biodynamic growers think that noxious weeds and dangerous pests can be repelled by spraying the plants or weeds with the burnt ashes of the weed seeds or animal remains (wild boars, rabbits, or starlings, for example, can all damage vineyards and their crops). This sends a message to the pests saying "don't come back here anymore." This is referred to as "ashing." Organic and particularly Biodynamic cultivators and vintners maintain that solar, lunar, planetary, and stellar cycles and rhythms can optimize vineyard and winery operations. This is due to the fact that biodynamic farming emphasizes life processes and forces that render vineyard soils susceptible to cosmic cycles and energies, in addition to using materials like composts in place of chemical fertilizers. However, some cycles allow for two weeks of optimal work time per month, such as pruning under a waning moon from full to new; other cycles, such as picking during a so-called "fire" constellation (Leo, Sagittarius, or Aries), which are all favorable for fruit crops like wine grapes, only allow for an average of one day of optimal work time per four. Larger vineyards that follow a celestial calendar must therefore make certain concessions.

Growing vines in a way that enhances the plant's inherent resistance to pests and diseases is the goal of organic and biodynamic vinegrowers. Being organic and biodynamic does not imply being "not sprayed." They achieve this by using organic composts and natural fertilizers (manure from animals frequently contributes to these), as well as misting their vines with pure "naturally occurring" chemicals and plant teas (tisanes) and liquid manures (purins), which are made by soaking different herbs, weeds, and plants in water. This limitation to naturally occurring chemicals typically forces vinegrowers to rely on products like sulfur, copper, specific oils, soft soaps, and commercially obtainable plant extracts (like rotenone, which comes from the Derris plant) that can be sprayed on vines. A tisane made from the weed known as

Common Horsetail (Equisetum arvense) is a commonly used spray that is believed to protect against fungal diseases. Some more contentious field preparations exist as well, like Horn Manure 500, which is claimed to encourage deeper vine rooting by promoting the formation of humus in the soil. This is created by burying fresh cow manure in the ground throughout the winter after stuffing it into a cow's horn. Similarly, Horn Silica is also utilized to enhance the vine's photoperiodic response. Another cow horn is filled with ground quartz (silica) and buried for six months to create this. Before applying the mixture to the soil or the crops, the contents of the horn are removed (after the ground has been dug up), diluted, and then stirred into water in a counter-rotating motion (a process called dynamising).

Biodynamic compost preparations need to be added to a compost pile in order for it to function properly. These are made from stinging nettles, ground oak bark, and the flowers of valerian, chamomile, dandelion, and yarrow (pressed to create a liquid extract). Some of these compost preparations are said to be "enhancing their capacity to sensitize the compost" and consequently the farm soil onto which it is spread by prefermenting animal organs like sheaths (red deer bladders, cattle intestines, animal skulls, and cow's peritoneum). It is understandable why skeptics frequently discuss the murky and enigmatic methods used by organic and biodynamic farmers.

Production Costs in Organic and Biodynamic Vineyards

Regardless of one's opinion regarding the efficacy of growing grapes in an organic or Biodynamic manner, it is undeniable that doing so raises production costs. Weed control is one of the main issues, and since conventional herbicides are essentially nonexistent, alternative solutions must be found. The mow-and-throw technique, in which grass and weeds mowed from the center of the alleyways are mechanically thrown underneath the vines to act as natural mulch, may help growers in warm and hot climates, especially those where there is little to no rainfall during the majority of the growing season, keep weeds down to an acceptable level.

Other mulches such as chipped bark, straw, and composted home waste can also be used; however, only organic growers are permitted to use these types of mulches. Cultivators, under-vine ploughs, and even weed-flamers may be employed in damper climates. If only because a tractor and driver must drive through the vineyard more frequently, these are typically thought to be more expensive methods of keeping vines free of weeds than using herbicides. Additional strategies include applying naturally occurring chemicals like soap, clove oil, or acetic acid, which can suppress weeds to some extent but require multiple treatments throughout the growing season to effectively control weeds. Some growers have experimented with sheep and goats, but the results are not always consistent when it comes to controlling summer weeds (as opposed to winter cleanup services). They will selectively graze and begin to nibble leaves and shoots unless they are confined to fairly small spaces. They also have irksome habits like rubbing up against posts and vines and breaking the weaker ones. In order to keep weeds under the vines and grass in the alleyways under control, "weed geese" have also been used; four per acre is recommended; you can eat or sell them afterwards.

Undoubtedly, pest and disease control in organic and Biodynamic vineyards is another area where costs are increased. Many of these slightly unusual methods of weed control look better in press releases and publicity shots than they actually perform in reality (though that's not to say that's a bad thing in and of itself). Growers will typically find that the natural chemicals and products they use are less effective than their non-organic counterparts, that they are contact rather than systemic, and that they require more frequent application in order to adequately protect the vines.

Additionally, their harvest intervals might be longer. In dry regions, this may not have much of an impact on fungal diseases, but it might make pest control especially challenging. When faced with a coordinated pest attack, natural insecticides like pyrethrins (made from chrysanthemum flowers) and barrier treatments like different oils and soaps may very well prove ineffective. Lower yields and fruit damage are what will happen. In comparison to grapes grown conventionally, crop quality is frequently lower in terms of pest and disease damage. This, along with the state of the fruiting buds and issues with disease during flowering, can all lower yields. For most organic and biodynamic vinegrowers, copper-containing sprays have been their primary defense against Downy Mildew; however, the amount of copper that can be sprayed has recently been lowered because copper lingers in soil and accumulates in the population of earthworms and other soil-borne organisms (slowly killing them). Now, it is only 4 kg for organic growers and only 3 kg for biodynamic growers annually, averaged over five years. This means that if an organic grower uses 5 kg in a year, they must cut back on their use in subsequent years to stay below the average of 4 kg per year. An attempt to spread out the average over seven years was made, but the EU turned it down.

Are biodynamic and organic wines superior to traditional ones? Since many organic and Biodynamic growers use chemicals in their vineyards—something that promotional materials and back labels frequently neglect to mention—if their wines have unique characteristics, it is not because they were grown without using any kind of non-natural intervention. It is difficult to find evidence that wines made from organic and biodynamic grapes taste better just because they are these practices. Though conventional growers can also choose to source from vines that are better tended and less heavily cropped, it's possible that this will enhance the flavor. They may originate from vines where there is a higher level of human input in terms of management and physical labor, and where manual labor-intensive procedures like canopy management, green-harvesting (crop thinning), and pre-harvest de-leafing are performed. It's interesting to note that, despite the fact that all of these methods limit yields, employ high-quality labor inputs, and frequently use animal manures, very few of the world's best wines are currently produced from organic or biodynamic grapes.

CHAPTER FOUR

THE VINE'S YEARLY CYCLE

Timings for the southern hemisphere can be calculated by adding or deducting roughly six months. Spring will inherently arrive earlier and all subsequent dates will be earlier in warmer climates. The harvest can begin as early as August in very early regions and in early years. In late years and in colder climates, regular harvesting will extend well into November. Although they always bear the vintage of the year they began the season, icewines can be harvested as late as February or March of the following year. Additionally, there will be some overlap in the timings, which will depend on the site, variety, and weather. Harvesting in the southern hemisphere begins in early January in very warm climates (and occasionally even in late December) and ends in mid-May in cooler climates.

January–February–March

The vine is dormant at the beginning of the year. The summertime green canes will have turned brown and become lignified, or woody, and their buds will have "ripened," or sealed against the winter. The vine goes dormant when the leaves have all fallen and the temperature drops. The plant will now have fixed any carbohydrates the vine produced, and their internal migration will have stopped. Then you can start pruning. Growers who have been busy in their wineries will actually hold off until after Christmas, once they have settled their wines. For growers in Europe, January 22nd, the feast day of St. Vincent, the patron saint of winegrowers, is considered a good day to begin pruning.

However, in many vineyards, pruning must begin as early as possible for purely manpower-related reasons, as pruning, along with cane removal and tying down of canes (in systems where canes are tied down), comprises a large portion of the winter work. The pressure to finish the task is reduced in spur-pruned systems where pre-pruning machines can be used and some pruning can be mechanized.

March/April

Well in advance of bud-burst, pruning should be finished, and all canes cleaned up and tied to the fruiting wires (in cane-pruned systems). This timing is only not applicable in cases where spring frost poses a risk, as previously indicated, and extra canes are left in case of frost damage, which can be removed once the threat of spring frost has passed.

All pruning, with the exception of pre-pruning, which will be covered in more detail later, is done using powered or manual secateurs. Electric, pneumatic, and hydraulic secateurs are fairly common and both speed up the pruning process and put far less strain on workers' hands and wrists than manual secateurs, as was already mentioned in the previous chapter. A saw might be required for larger cuts, but powered secateurs can handle wood up to 50 mm in diameter.

Additionally, a range of powered pruning gantries and rigs are available, allowing pruners to be protected from the weather while they work and, more importantly, to be serenaded with music while they do so. It's not easy to stand around in 150 mm (6") of snow in the middle of winter and try to stay positive and focused on the task at hand. Small wheeled seating platforms, often battery powered, that allow pruners to sit and move themselves along the rows are a common sight in France for very low-trained vines. Depending on whether the vines are spur-pruned or cane-pruned, the pruning procedure itself can be divided into four or three parts.

The first step in cane pruning is choosing the fruiting canes that will be needed for the next year. Next comes cleaning and trimming the canes to the proper length. Third is "pulling out," or chopping off unwanted wood from the trelliswork. Lastly, the pruning process is finished by tying the fruiting canes to the wires, also known as "tying down." Although in many vineyards the selection is done according to a fairly strict formula that can be repeated from vine to vine, picking the fruiting canes requires some expertise. The wood to be chosen needs to come from a point on the trunk that guarantees that the vine crown stays in the proper location. This point is typically below the fruiting wire, which is the lowest wire, and must be positioned to provide the maximum amount of fruiting area. The area of fruiting space on the vine will decrease if the crown grows too high because the canes will get shorter. Occasionally, when dealing with older vines, one must take rather drastic measures, such as pruning off older wood and starting anew from a lower position on the vine that approaches the ground. Anti-fungal paint is frequently applied to large (and occasionally smaller) pruning wounds to prevent the growth of diseases like Eutypa lata and Esca. It will be clear which canes are to be kept once the fruiting canes have been selected by removing any unwelcome canes that surround them.

In many vineyards, less specialized workers will follow behind the more experienced staff members to finish cutting out and removing the remaining undesired wood and tidying up the canes that will be kept. The unwanted wood is typically placed in the middle of the rows in vineyards and later ground up (mulched) by a tractor-mounted flail mower or pulverizer. As an alternative, the wood can be packed and taken out of the vineyard to be burned or composted. Additionally, I have witnessed devices that can effectively vacuum up wood, spitting it out the back in compressed logs that are ready to be placed directly onto a barbecue. To burn the prunings in place and keep the pruners warm, some French vineyards still use chariots de feu, which are essentially half an oil drum that have been perforated with a few holes and welded onto an old wheelbarrow frame. However, these are getting harder to find.

The annual canes are bent over and/or tied down as the last step in the cane-pruning process. Canes are trained over one wire and fastened to a lower wire in bent-cane systems (many varieties of Guyot, Pendlebogen, etc.). You can leave the canes bent like this until the sap begins to flow through the vine and makes the canes more pliable. The canes are fastened to the wires using a range of materials, from straightforward paper-covered wire twists to semi-automatic tying systems that employ plastic tape, wire, or cord.

Pollarded willows are grown alongside vineyards in some regions of Europe. The thin red or yellow shoots that these willows produce is harvested, soaked in water to make them pliable, and then used for this purpose. In vineyards with spur pruning, the vines can be pre-pruned with a mechanical device attached to a tractor. This device will snip off most of the annual wood and break it into small pieces that will fall to the ground. The only manual pruning that will remain will be picking out the shoots that will become spurs and cutting them back to the proper length. Spur-pruned systems, with the exception of combination systems like Sylvoz pruning, do not require bending over or tying down, and this, along with the ability to perform pre-pruning, makes them significantly less expensive to operate. While growers in areas where spring frosts can be an issue will try to postpone pruning as long as possible, all pruning should ideally be done well before the sap begins to rise. By pruning late, frost damage can be minimized by slightly delaying bud-burst. At this point, other tasks in the vineyard might include fixing trelliswork, planting new vines in areas where spring frost is not a concern, and spreading fertilizer and manure when the ground is dry enough for tractors to drive over (though these tasks can also be completed when the ground is frozen).

April–May: Several things begin to happen in the vineyard as the weather warms and the temperature of the soil rises. The vine emerges from dormancy when the cut ends of the canes begin to bleed (produce sap). Although one would assume that a vine producing a lot of sap would somehow weaken it, studies have shown that this is not the case. Young shoots will eventually emerge and begin to grow as the winter buds begin to enlarge and turn woolly. We call this bud-burst. Wet soils will warm more slowly, and in colder climates, increases in soil temperature are crucial for encouraging bud-burst (soils at 200 mm depth must rise to 8°C to 10°C before vines begin to grow). Weeds and grasses will begin to grow on the vineyard floor and will need to be removed.

Some bud-rubbing or shoot-removal will be necessary in many vineyards, particularly those with yield restrictions or where past experience has shown that there is usually an excess of shoots. Eliminating extra buds or shoots will make room for the remaining ones to grow in more exposed areas, reduce competition, and encourage the development of fruit buds on the canes that will bear fruit the following year. This kind of shoot removal is less common in cane-pruned vineyards and more common in premium spur-pruned ones. The area immediately below the vines and the space between the rows (the alleyways) are the two areas that require different levels of maintenance on the vineyard's surface. It is normal to keep the soil free of grass or weeds directly beneath the vines and in a band of about 250 mm to 300 mm either side of the vines' center line (so about 500–600 mm overall).

Aside from the desire to prevent tall grasses and weeds from growing up and interfering with the crop, keeping the soil free of obstructions reduces the likelihood of fungal diseases developing by removing competition for nutrients and moisture from the immediate vicinity of the vine's roots.

Weed Control

There are several other ways to maintain the vineyard alleyways: you can use disc, spring-tine, or powered harrows to cultivate them; you can mow grass over them; you can plant certain plants (known as "green manuring") that yield valuable amounts of green matter that can be chopped and re-incorporated into the soil to raise the humus levels; or you can use herbicides to keep the alleyways free of weeds. Which path to choose will be determined by practical considerations, regional customs, and the grower's overall philosophy.

One of the challenges associated with cultivation is that the cultivated soil may not always be suitable for the passage of heavy harvesting machinery such as tractors and pickers during wet harvests. Sometimes, growers will only cultivate every other row, leaving the uncultivated rows for picking machines or grape-trailers. While green manuring and mowing are easy and efficient techniques, they may cause the vine to lose moisture or use more expensive irrigation water in arid regions. Because of the justifiable desire to minimize the use of herbicides, the practice of keeping alleyways clear has become less common than it formerly was.

Nonetheless, herbicides are typically applied beneath the vines if they are intended to keep alleyways free of weeds. This method is referred to as "total herbicide." Throughout the entire growing season, the soil beneath the vines needs to be kept free of debris, and the alleyways need to be maintained, regardless of the techniques employed. The issue of managing pests and diseases emerges when the vine begins to grow and new, young leaves begin to appear. Early spraying is typically required in vineyards to combat diseases like Oidium and Downy mildew, with botrytis control beginning a little later in the growing season. Throughout the growing season, pest and disease control will be maintained (though fewer sprays will be applied during dry seasons), and it will only conclude when the harvest is imminent. Every chemical used in agriculture is subject to legal approval before being used, and each chemical has a harvest interval that varies depending on the material. In addition to this legal requirement, many chemicals should be avoided if you want quick fermentations because they will negatively impact yeast growth in the must if they are left on the grapes. In areas and locations where, past experiences have indicated that spring frost poses a risk—early spring is the most likely time to occur and right after—proactive steps are frequently implemented to safeguard the vines and their upcoming fruit.

June–July

When annual shoots grow, most of them will grow up between the wires in many pruning systems (especially VSP systems, whether cane or spur-pruned), and their tendrils will support and guide the canes as they grow. Certain shoots will inevitably fall outside the wires and require tucking in. There are several ways to deal with this issue: spreader arms that keep the catch-wires apart and aid in capturing the growing shoots; movable catch-wires that can be moved up to trap the foliage as it grows; and even machines that do the entire task mechanically. Tucking-in is not necessary in certain spur-pruned systems (GDC, Blondin, Sylvoz, etc.) where the annual growth is left to hang down from the winter canes naturally. But even with these systems, there's a chance that some cane positioning will still be required to open up the leaf-wall's centers and let light through the canopy. The vine will typically begin to flower during this time. The flower caps will break off, the flower clusters will open, viable pollen will be generated, and pollination (or fertilisation) will occur once the annual shoots have reached a length of, at most, one meter. A robust vine with fecund flowers, warm, dry weather, and ideally a light breeze to facilitate pollen migration from the stamens down the style to the ovary (the female portion of the flower) are all necessary for successful flowering. Though they frequently visit vine flowers during pollination, insects such as bees, beetles, and flies are not thought to be essential for successful flowering. Similarly, since there is very little—just a few millimeters—between the stamens and the ovary, wind has no effect on the pollination of hermaphrodite vine varieties.

The typical five-stamened vine flower has 20,000 grains of pollen in it, which is far more than what is required for pollination to be successful. Despite this, the flower can scent the vineyard with a distinct light citrus and honey scent. During this time, a cold, rainy spell will hinder the growth of pollen and its transfer from stamen to ovary, preventing the flower caps from fully opening. This is the most critical period in the growth cycle for the upcoming crop: an early, rapid, even, and successful flowering will produce a full crop that ripens evenly, making harvest date prediction easier; a cold, damp, delayed, and extended flowering will produce a lower, or nonexistent, crop that ripens unevenly, resulting in grapes that arrive at the winery at varying levels of ripeness. The use of picking machines, which are unable to distinguish between ripe and less-ripe bunches, exacerbates the issue of an extended flowering that leads to uneven pollination and, consequently, uneven ripening.

Furthermore, incompletely shed flower caps from flowering under unfavorable conditions increase the risk of bunch rot in varieties with tightly packed petals and incompletely shed flower caps that become embedded within the bunch. The quality of the flowers, which is influenced by the growth circumstances in the year before, is a major factor in the success of flowering. Between 15 and 17 months prior to harvest, the flowers undergo what is referred to as their "initiation." That is, for a crop that is harvested in October, in May or July of the previous year. When flowers form, a variety of factors come into play. These include heat and light levels, the overall health of the vine, whether or not it is under stress, and mineral imbalances. All of these factors can affect whether or not a vine receives an adequate supply of viable flowers and pollen. Low pollen viability levels are also present in some varieties and some of their clones, which could be the cause of subpar crops. Warm, dry, still conditions are ideal for pollen transfer to the ovary, and good flowering is also highly dependent on the weather. Flowers occasionally open at about 15°C; normally, they open at 17°C, and quickly at 20–25°C. Humidity and/or moisture will inhibit pollen growth and reduce the certainty of pollen transfer from stamen to ovary, which will almost certainly result in coulure and/or millerandage60. Some growers trim (or tip) their vines' shoots during flowering to aid in the process of flowering. Numerous studies have demonstrated that pruning the shoots at this point, when flowering is between 30 and 50 percent finished, produces larger crops and can increase yield by up to 25 percent when compared to untrimmed vines.

Girdling is another method used in many table grape vineyards to increase yields. This involves cutting off a small ring of bark from the fruit-bearing arm just before flowering. By stopping specific nutrients and hormones from the roots from reaching the flowers, this effectively deceives the flowers into more effective pollination. There is no long-term harm, and the bark circle quickly regrows. Additionally, three weeks prior to flowering, table grape vineyards can benefit from spraying their vines with gibberellins, which reduces bunch compactness and increases the size of individual grapes—a feature that makes table grapes desirable. A growth retardant based on the chemical chlormequat (also referred to as Cycocel or CCC), which decreases shoot elongation and has an impact akin to shoot tipping, is another chemical used to aid in flowering. A product known as Regalis Plus may be sprayed during flowering time in some areas in order to "improve bunch architecture," or open up the bunch, in order to help control botrytis.

Some growers also choose to go through their vines right after flowering, emptying their sprayers of liquid while keeping the air fan running to remove any remaining stamens and caps from the bloom. Numerous studies have demonstrated that Botrytis frequently begins on the flower debris that is caught inside the bunch. It becomes impossible to apply fungicide to the center of the bunch as it grows and approaches "bunch close." Many growers will begin both leaf removal and vine trimming (hedging) as soon as flowering is finished and the tiny berries begin to swell and expand to the size of peppercorns.

Summer pruning, or the act of trimming after flowering, is a common practice in vineyards to control foliage, remove unruly side shoots, and preserve an open, thin canopy. By doing this, the alleyways will remain open for machinery and tractors to pass through without uprooting bunch-carrying shoots that have fallen from the canopy. Until they enlarge to 50–80% of their maximum size, the vine's leaves are net consumers of carbohydrates; after that, they begin to add to the plant's sugar buildup. When they get to their largest size, they no longer contribute as much to the sugar buildup.

Therefore, pruning must be done with caution, removing only enough foliage to open up the canopy and allow light, air, and pesticide sprays to reach the center of the leaf-wall. Tipping may lead to more side shoots (lateral shoots) forming in vineyards with moderate to high vigor. While these may help with sugar accumulation, they will also cause overcrowding and shading within the leaf wall. The removal of leaves, which can be done by hand or by machine, usually begins around the time the first trimming is done. Usually, growers will remove one-third of the fruiting zone leaves during the first pass, a second third during the second pass, and a final third during the last pass. Naturally, some leaves will regrow in between passes, leaving the vine partially bare throughout the fruiting zone. Only the shaded side of leaves may be removed in warm to hot climates to avoid sunburning the fruit. Throughout these months, weed control, alleyway management, and pest and disease prevention will all continue.

July–August

The vine's growth will slow down and shoot extension will not be as rapid during these months, which are the hottest in the northern hemisphere's summer. Nonetheless, summer pruning will persist and certain shoots, particularly side shoots, may very well continue to grow. In order to aid in the development of the grape sugars and to enable pest and disease treatments to reach the bunches, the goal of canopy management at this stage is to open up the leaf-wall, allowing light, air, and heat into the center of the canopy. Additionally, this is véraison, the time of year when red vine varieties' grapes begin to turn from green to red and the ripening process truly begins, with sugars building up in the fruit and acids beginning to decrease. This is the point where "green harvesting" can be done. The goal of green harvesting, also known as bunch or cluster thinning, is to lower the total amount of the crop while also removing the least developed grapes from the vine. When it comes to red wine varieties, véraison is the best time to do this because the grapes that are least ripe can be easily identified by their color change. After that, these can be removed or severed and thrown away. The crop will be reduced by bunch-thinning at this point, but the overall effect might not be as noticeable as it first appears. Removing a third or half of the bunches numerically does not reduce the crop by anything close to the same percentage because the bunches removed are usually much smaller than the bunches retained, which are the cane's third and fourth, if there are any.

Furthermore, the vine will try to make up for the lost crop by growing the remaining bunches and berries in size—that is, if it receives an adequate amount of water. This not only undermines the attempt at crop reduction, but it may also lower the wine's quality, particularly in the case of red grapes where tannin structure and wine color are largely determined by the skin to juice ratio. Removing all but one bunch per cane is almost always necessary to achieve a really significant reduction in yield, something that most growers find extremely difficult to undertake. However, there will be uneven levels of ripeness between the first, second, and third bunches on the cane in years when the flowering process has been long and drawn out, usually due to cool, wet, and windy conditions at flowering. Green harvesting does, at least, get rid of those bunches that will ripen last. When vines are harvested by machines, green harvesting could lead to better quality because the machines aren't able to tell the difference between ripe and unripe bunches.

However, there are additional costs associated with green harvesting. With the exception of early regions or very early years (Europe in 2003, for example), August will be a period of decreased vineyard activity. It's a good idea to keep weeds under control, trim and open the canopy, and minimize pest and disease damage. This will allow vetirons to rest before harvest time. In Europe, September through October is when most harvests take place. It was once believed that there was a 100-day gap between the start of flowering and harvest, as was previously discussed in the climate section of this book, but most of the time this is no longer the case. The period of time between flowering and harvest can range from 135 to 145 days in many regions of the New World.

Undoubtedly, increased control over pests and diseases—particularly Botrytis—as a result of improved canopy management, more effective fungicides, and better application techniques has given growers the confidence to let their grapes hang longer. One must also take into account the use of harvesting machines in many vineyards, which can pick in an hour what five people could have done in a day, and the desire to produce better wine with greater extract, higher alcohols, and riper flavors.

Additionally, machine harvesters have the ability to work around the clock, which is useful in areas with hot summers. In these areas, grape harvesting typically begins well in advance of first light and lasts until temperatures rise too high. The precise date of harvest will depend on a number of variables, including how well-developed the flavors and colors are, how well-suited the grapes are to the type and quality of wine that will be needed, how much the sugars have risen, and how high or low the acids have gone depending on the style of wine. Other variables that aren't directly related to the grapes' actual ripeness level include practical issues like whether pickers can be hired, whether a picking machine is available, whether the winery has enough tank space, what the weather is going to be like, whether a weekend or holiday is approaching, and whether grandma's birthday falls on the weekend. There is life after the winery for winemakers! Typically, growers test their grapes for pH, sugars, and acids three to four weeks before harvest. However, an experienced winegrower can also determine the approximate date of harvest based on factors like leaf color, berry color change, pip condition, and grape flavor. A lot of winemakers will walk around their vine blocks, sampling individual grapes to get a sense of how the ripening process is going. In relation to a harvest date, many winemakers will also talk about the "physiological ripeness" of their grapes.

The term "physiological ripeness" is debatable, and there's no surefire way to pinpoint the precise moment this occurs. A variety of factors are considered, including skin color and condition, pulp and juice flavor, pip color and ripeness level (bitterness), and a general sense of when the grapes are at their best to produce the best wine possible. In warm and humid regions, grapes that are harvested at their physiological ripeness point may have too much potential alcohol and too little natural acidity; these factors may need to be adjusted if the finished wine is to be properly balanced. A lot relies on the preferred wine style.

Naturally, in many appellations, the decision about when to harvest is left to a local committee made up of growers, winemakers, and state representatives. They set a date for each region, and occasionally even each sub-region, so the winemaker is relieved of their final say in the winegrower's year. The physical state of the grapes is another important factor to take into account when choosing a picking date. Have they sustained any damage, and what is the likelihood of an invasion by Botrytis or another fungal disease? With white grapes, skin finish and condition are not as important as they are with red grapes, where all of the grape's components— aside from the stalks—end up in the fermentation tank. However, grapes damaged by botrytis may lose a significant amount of their fruit and freshness and even give the wine a moldy taint.

Some level of bird protection will be required in some areas, particularly in remote vineyards with adjacent woodland areas where birds can roost. Static methods like firecrackers hung in strategic places, automated gas-guns that make a loud bang, and various visual and audible warning devices (the distress calls of the species of bird doing the damage are quite popular) will help scare off flocks of birds. Regular bird patrols with gunfire and general disturbance will also deter birds. Seeking to attract raptors to build their nests in the area is also a smart move, and some grape growers have even utilized raptors piloted by skilled falconers to rescue their crop. What works well one year might not work well the next, and if birds are an issue, it usually sparks a brief conflict that ends with the harvest. The yearly battle against birds has led to a significantly decreased – in many cases nonexistent – bird population in regions of the world where vineyards have been around for a long time and where vines are typically the only crop grown there. The issue has also been largely resolved.

Harvest

However, overall netting might be the only option in more recent areas. Netting is costly and time-consuming, but it works; in New Zealand, for example, about half of all vines are covered with nets to keep birds out. Vineyards situated close to other fruit crops, particularly those that have been harvested already (like pears or apples), appear to sustain significantly less damage from birds because the birds will eat the riper fruit that has fallen rather than the relatively acidic, unripe grapes. Bird damage can result in a reduction in yield, but pecked fruit also becomes susceptible to bunch rots of different kinds. A vineyard with a bird problem needs to find another way to protect the grapes; early harvesting is undoubtedly not the solution.

November - December

There is much less work to be done in the vineyard after the harvest is finished. Certain cultivators may use a clean-up herbicide to eradicate any weeds that have sprouted in the latter half of the growing season. Sometimes a winter-wash of a fungicide with a copper base is used to promote the canes' hardening and ripening. Vineyard cultivators may choose to perform a final clean-up cultivation to allow the winter rains to seep into the subsoil, either beneath the vines or in the alleyways (or both). In certain areas, it's customary to till the ground near the vines' base, or "buttage" in French, piling soil up against the grafts to shield them from damage caused by winter frost. (The opposite procedure, known as debuttage, will occur in the spring to remove the dirt.) When the vines are fully dormant in spur-pruned vineyards, pre-pruning machines can make a first pass through them; in warm climates, proper pruning can then begin. In warm climates with low risk of winter frost, new vines may be planted in addition to routine maintenance like post replacement and repair. However, these are the quiet months in many vineyards, when work is focused in the winery and not much else happens.

Does quality and yields have a relationship? One of the most widely held beliefs—which you will undoubtedly hear reiterated—is that yield and quality are inevitably inversely related. Put another way, wine quality will always increase if the yield is decreased. It's true that in many vineyards, a lower yield will result in higher-quality wine because, all else being equal, a single vine will have an easier time ripening a smaller amount of fruit. But is that the only way to improve the quality of wine? Maybe those vines could have ripened the original crop to the same degree and produced an equally good (and more plentiful) wine if the vineyard had undergone different pruning and/or training system adjustments, and if canopy management practices (bud-rubbing, shoot-positioning, summer pruning, leaf removal, and possibly a light bunch-thinning) had been implemented with a little more care. For the type of grapes they are attempting to produce, the majority of growers will have an idea of what crop level their vines will support.

Growers who are bound by fixed-price contracts or who choose to take their chances on the annual spot market may believe that the safest approach to optimize their earnings is to invest as little as possible in crop growth and, while adhering to any yield, sugar, and acidity requirements imposed by any contracts they may have signed, to select as large a weight of grapes as possible. In reality, growers who try to predict their yields in order to keep their crops below the set levels but are limited by appellation regulations to a specific yield are up against an extremely challenging task. Test picking, which involves selecting about ten vines from a vineyard (a task that most growers find extremely difficult), can be used to estimate yield.

The average yield per vine can then be multiplied by the total number of bearing vines in the vineyard, keeping in mind that many vineyards, particularly older ones, will have a percentage of underperforming or even missing vines. In practice, this means that growers will inevitably modify their picking techniques once they begin and get a sense of the actual weight of crop hanging on their vines. Light cropping vineyards (those below the appellation limits) will be plowed of their last bunches; those that appear to be yielding more will be picked selectively; the third bunch, which is typically the smallest and least ripe, will be left for the birds to harvest, or perhaps for a nearby grower to pick from a young, low-yielding vineyard.

Selecting grapes up to the appellation yield level and discarding the remainder essentially controls the wine market and adds very little value to the wine. Young vines that are well-watered and have the same crop level as their older neighbors will yield fruit that is either as good or worse. This is a crucial point to remember. Many winegrowers will tell you that vines will produce excellent quality fruit in their first one or two crops, even as early as their second or third year of growth, free from diseases that haven't had time to establish themselves in the vineyard. However, keep in mind that this is only true at low cropping levels. When everything else is equal, level of yield is the single most important component of quality. When vines reach the age of twenty or more, they typically become less vigorous, yield fewer fruits, are better balanced, and have more exposed fruit. Better wine from older vines is a direct result of this factor alone.

CHAPTER FIVE

GRAPEVINE DISEASES

Like any other intensive agricultural or horticultural crop, vines can be harmed by a wide range of diseases, pests, viruses, and other viticultural issues. If these issues are not addressed, it can be challenging to produce crops that are profitable. However, a number of factors have simplified vine care in recent decades, and vineyard owners and managers now have a stronger arsenal of defenses against the frequent issues. Both rootstock and scion varieties of clones are superior than ever, and when choosing material for cloning, virus resistance and disease resistance rank highly.

Grafted vine quality has increased since the early 1990s when the EU implemented a plant passport system, which included concurrent source inspections and virus testing. Although as the incidence of trunk diseases shows, hardwood diseases have by no means vanished, better hygiene in rootstock vineyards, scion mother-gardens, and vine nurseries has also reduced the likelihood of viruses and diseases being transmitted to destination vineyards. But nothing in nature stays still for long, as evidenced by the 2008 discovery in California of what was initially believed to be a leafroll virus variant but turned out to be a new virus known as Red Blotch, indicating that vineyard inspections still require caution. A large portion of vineyard work can now be done from the seat of a tractor thanks to improved trellising and training techniques that are better suited for mechanization. Regular spraying can also be completed quickly and effectively in areas where it is necessary. The majority of new tractors come equipped with air-conditioned comfort-cabs and four-wheel drive, making them more sophisticated than ever.

On fruit farms and vineyards, GPS-guided tractors—both manned and unmanned—are beginning to appear. These tractors are common in horticulture and arable land. Modern spraying equipment, like nozzles that create uniformly sized, much smaller droplets, is a huge improvement over the outdated "spray to run-off" kind of sprayers that were once in use. More hectares can be sprayed for any given tank size thanks to low volume and recirculation sprayers, which catch, filter, and recirculate spray that misses the target. This significantly increases productivity. Recirculation sprayers use up to 50% less spray than conventional sprayers, cut chemical costs in half, and produce less environmental waste. A double-sided recirculation sprayer in use in a UK vineyard.

Additionally, there is a growing recognition that managing the canopy can be crucial in combating illnesses. Fungal disease is fought off by heat, light, and air. If the vineyard canopy is open, with no more than three to four leaf layers and adequate air drainage through and especially beneath it, the vines will dry out more quickly and fungus issues will be minimized. Tools that help with timely canopy management, like leaf strippers and mechanical trimmers, are now standard.

Numerous factors, such as the vineyard's location in the world, climate, and specific weather of the year, as well as the variety, clone, and rootstock being grown, will affect the range of pests, diseases, and other viticultural issues that an owner must deal with. Along with wine style and quality, economics will also be important. It goes without saying that every grower wants to deliver 100% clean, unharmed grapes to the winery, but occasionally there is a trade-off between expense and benefit, and production economics must be taken into account. Because machine harvesters are being used more often, growers must have extremely clean crops in order to avoid being penalized at the winery door. You can instruct your pickers to only pick clean grapes, but a machine harvester will not be very successful with such instructions! There will be significant yearly variations as well, which are influenced by heat, sunshine, and rainfall.

Certain illnesses are spread by insects, or "vectors," and managing these insects—using pesticides and reducing their habitat—may help manage diseases that don't seem to be related. A number of other variables will also come into play, including exposure to the dominant winds, vineyard floor treatments, and the presence of natural predators (in the case of pests). Today, there is also a greater focus on monitoring to determine the pressure from pests and diseases. Targeted traps are beginning to appear in vineyards; some of them are equipped with remote cameras that log the number of times the trap is visited in a given day. In vineyards, small weather stations that capture a variety of data are also becoming increasingly prevalent. The information gathered from these can then be utilized to forecast various "disease events" and schedule sprays appropriately. In the past 25 years, a number of factors have caused farmers and growers to abandon the "spray first, ask questions later" approach and instead attempt to understand the causes of problems, take all reasonable steps to mitigate them using non-invasive, non-chemical methods, and, in the last resort, use as little chemical as possible to protect their crops. Naturally, this is not entirely selfless. Chemicals are expensive to purchase and use; in terms of production costs, they usually come in second (behind labor), and their use is strictly regulated by law. The public has also become much more aware of the need to steer clear of over-sprayed food items, and farmers and growers now recognize that they are stewards of the land they cultivate and that sustainability is an essential component of their business practices. In addition, the organic and biodynamic farming movements have contributed to raising awareness of the possibility of farming with lower chemical inputs.

All of these considerations have forced those involved in the management of pests, diseases, and weeds to consider methods of reducing chemical inputs for the good of the land, the environment, wildlife, the general public, and farmers and growers. Fight raisonée, also known as lutte intégrale, is the French term for Integrated Pest Management (IPM), which has been practiced for several decades. The goals are essentially the same no matter where it is applied: identifying the root cause of the issue, keeping an eye on external factors like the weather to determine when the issue is most likely to occur, and acting quickly to prevent problems rather than waiting for more expensive, curative solutions. IPM also makes use of natural predators whenever possible, such as the strobilurin-based anti-Oidium treatment or parasitic wasps whose larvae kill specific caterpillars.

Some issues, the most well-known of which is probably Pierce's Diseases, have proven to be uncontrollable through chemical means; hopefully, this one will eventually be resolved with the introduction of a natural predator to the leafhopper insects responsible for spreading the bacterial disease: the grey-green and glassy-winged sharpshooters. Similarly, researchers think they might eventually discover a helpful nematode that would eliminate Phylloxera larvae below ground, negating the need for grafting in certain soil types. Reducing soil erosion on sloping sites and increasing the number of naturally occurring predators in the vineyard are two benefits of using appropriate cover crops planted in the middle of the rows. It might, however, also expand the habitat available to disease vectors.

On the other hand, many small growers find it challenging to fully implement IPM. Even though their chemical suppliers aren't the most impartial advisors, small growers still rely on their recommendations for spraying, but they always try to cut back on the concentrations they use, if only to save money. Growers who are part of cooperatives or have contracts with wineries to supply grapes are frequently restricted in their options because they must adhere to the spray schedules set forth by the grape's final consumer. Nonetheless, every grower will adjust their chemical inputs based on their knowledge of their own vineyards and varieties, and the desire to save money is a strong motivator to cut back on material usage. Nowadays, many growers have small weather stations that provide real-time information about the conditions in their vineyards. The majority of these stations are connected to a reporting system that provides disease predictions based on temperature, humidity, and rainfall, enabling the growers to more precisely time their spraying.

Having said all of this, there is a sneaking suspicion that some growers are joining sustainability programs more out of obligation to present well to trade wine buyers and the general public than out of desire.

Harvest Intervals

Every pesticide used in agriculture must pass stringent testing before being approved for use. Chemicals are approved for use either "off-label" or "on-label." Whether the product has been classified for use on a broad range of crops and circumstances (on-label) or for minor crops and specific restricted circumstances (off label) will determine this. This approval will specify the product's intended use, the kind of sprayer to use it in, the number of times per growing season and per application, the amount of product that can be used per application and per season, and whether or not it can be tank mixed with other products. Last but not least, the approval for fungicides and insecticides will also specify the "harvest interval," or the number of days that must pass between the last treatment and harvest.

Naturally, it will vary from crop to crop and product to product. The majority of products used on wine grapes have fairly long harvest intervals (up to 56 days for some products), as there is a risk that traces remaining on or in the grapes will prevent yeasts from multiplying and affect fermentation. This is in contrast to grapes used on juice, the table, or for drying. There is a Maximum Residue Level (MRL) for every pesticide that should not be surpassed. Hence, before applying their final spray—which is usually intended to combat late-season diseases like Botrytis—growers must assess if they can allow the crop to stand for the two to three weeks leading up to harvest. The accuracy, affordability, and ease of use of modern residue testing equipment, which is based on GC-MS (gas chromatography-mass spectrometry), has increased to the point where it can now detect residues as small as 0.01 mg/kg, or, as one prominent wine analyst put it, "one grain of salt in 100,000 litres of wine," which is far below all MRLs. Growers may still produce grapes and wines with noticeable (but officially non-harmful) pesticide residues even if they are using the right tools, the right concentrations, and paying close attention to harvest intervals.

Roses and Vines

Roses are frequently planted at the ends of rows of vines. Although they serve only as decoration now, in the past they served as a warning system for vine mildew. The same meteorological conditions that led to the development of mildew on roses—which, incidentally, was unique to roses and did not spread to vines—would also result in the development of mildew infections on vines, but 10–14 days beforehand. Then, vinegrowers would be aware that their vines needed to be sprayed. Many contemporary rose cultivars have undergone clonal selection, hybridization, and cross-breeding to reduce their susceptibility to diseases like black spot and mildew. Diseases caused by fungi, bacteria, and organisms resembling bacteria: In the world, there are countless varieties of bacterial diseases that impact vines.

The diseases that are most common in terms of geography and economic harm are Botrytis, Downy Mildew, and Oidium, which are ranked top. Botrytis (Botrytis cinerea) is a very common fungus that attacks a wide variety of vegetable, salad, and fruit crops. It is also known by several names, including Pourriture Gris, Grey Mould, Grey Rot, Bunch Rot, Sour Rot, and Stein Rot. Botrytis is attacking lettuce that turns brown and begins to liquefy, as well as strawberries and raspberries that shrivel up into a mass of gray or mauve mold. It prefers moist, humid environments and grows best in areas with sugar, which is why it enjoys grapes, especially in regions with summer and fall rainfall.

Nurserymen storing wood for grafting and table grape growers who must store their grapes also face this issue. Botrytis not only causes crop loss on a physical level, but it also lowers crop quality. A wine without noticeable Botrytis taint may be produced from heavily infected white grapes if additional Sulphur(iv)oxide is added to the picking bins, pre-fermentation settling is improved (especially with bentonite added), and fermentation is accelerated with a high dose of active yeast. However, some fruit flavors and aromas will have been lost, the color may be slightly darker, and more Sulphur(iv)oxide will be needed in the finished wine to achieve a good level of stability.

When red wines undergo fermentation with skins and pulp present, they can develop more severe issues like moldy taints and off-flavors that require charcoal fining to get rid of. In vineyards that are susceptible, botrytis is typically present all year long. It overwinters as hard, dark-brown encrustations known as sclerotia, which are found on old wood and on unpicked or fallen grapes that have fallen to the vineyard floor. When the conditions are right in the spring, the sclerotia will supply the fungus's nucleus, allowing it to infect the vineyard's young green tissue.

Spraying against will begin when the first one or two leaves expand to about 25 mm across in vineyards in regions with summer rainfall and cultivating varieties known to be susceptible. This may well continue at intervals of ten to fourteen days until shortly before harvest. Certain years will require two or three pre-flowering sprays, possibly one during flowering, and two to three after; other years, particularly in dry growing conditions and when the variety in question has some inherent resistance, may require only two sprays before flowering and two after.

Like many diseases, if left unchecked early in the season, it can cause damage that makes life much more difficult later on. Botrytis will infect the flowers, and in a year with wet flowering conditions, the disease may stay trapped in the center of the bunch, particularly when the flower caps are struggling to separate from the flowers. The disease will be deep within the grapes as they enlarge and the bunch closes, making it impossible to spray to the center, though systemic spraying will be beneficial. Any remaining flower debris in the closed bunch will cause classic bunch rot, which could cause a significant loss of crop. When flowering ends, growers in vulnerable areas typically make sure to give their vines a thorough blast with a high-volume air-assisted sprayer that is loaded with an efficient anti-Botrytis chemical. This should accomplish two goals: it will hopefully blow out and scatter any remaining flower cap and stamen fragments within the bunch and cover the remaining portion with a protective spray.

Another disease with a reputation for developing a resistance to specific chemicals is Botrytis. The primary anti-Borytis products of the 1960s and early 1970s contained benzimidazole, also known as Benlate or Benomyl, an active ingredient that proved to be highly efficacious for a long time. But after applying them to the same fields for a number of years, the growers noticed that not only did their worm populations begin to decline, but the products' efficacy also decreased. It was then discovered that their specific strain of Botrytis, which has a tendency to become site specific, had developed resistance. Rovral WG, which has a 14-day half-life, can still be used in conjunction with other products up to a maximum of four applications per season.

The next generation of anti-Botrytis products were based on active ingredients called dicarboximides; Ronilan, Rovral WG, and Sumisclex were the three most well-known. These proved reliable for a decade or more before, once again, their effectiveness tailed off and resistant strains of Botrytis emerged. A brand-new product named Scala, which was based on the active component pyrimethanil, was released in 1995. The use of this systemic product is limited to a maximum of two applications annually (all prior products were contact only). Though it only has a 7-day HI outside of Europe, it is one of the most effective materials against Botrytis, with a 21-day half-life in the UK. Other products with different active ingredients, like Prolectus, Teldor, and Switch (which can only be used once or twice a season), can also be used to treat Botrytis. In the UK, three more products are approved for use against Botrytis on grapes in addition to the pesticides mentioned above. Despite being one of the priciest products on the market, Serenade ASO is a bio-fungicide that is based on Bacillus subtilis and has no HI, making it useful for the final anti-Botrytis spray. Another bio-fungicide with a zero-day HI is called Prestop, and it is based on Gliocladium catenulatum.

Although not everyone who has tried these bio-products has been completely satisfied, it is unclear how effective they are. The third anti-Botrytis suggestion is to use Regalis Plus, a growth regulator. This spray will open up the bunch by lowering the bunch's berry count if applied near the end of the flowering period. This might be taken into consideration in years with favorable flowering conditions, where large bunches are anticipated, particularly with tightly-bunched varieties (Chardonnay and some Pinot Noir clones, for example).

In order to prevent the development of resistant strains of Botrytis, wise growers typically employ a variety of products during the growing season. Scala has proven to be a useful tool in the winegrower's toolbox to this day. Only in cases where your grapes stay healthy until picked can you really achieve an extended hang time. Additionally, a secondary invader, botrytis will feed off of grapes that have been harmed by hail, insects, machinery, or birds. In these cases, where the grape's skin has split and the pulp is visible, botrytis will grow and eventually spread throughout the entire bunch. It may also simply damage the bunch's stem, resulting in "stem rot," which will cause a significant number of bunches—possibly even a very significant number—to fall to the ground. This may be especially problematic in years with heavy yields, as a general depletion of the vine's resources may have weakened the stems. Nonetheless, there are other methods of controlling Botrytis besides chemical spraying; appropriate canopy management, well-executed trellising, and pruning all assist.

The development and spread of the disease can be slowed down by an open leaf canopy, some shoot and leaf removal in the fruiting zone, and adequate air movement beneath the vines. Spreaders for foliage can also be used to divide the bunches based on the canes. Water dripping from leaves and shoots quickly spreads spores, so the faster the leaves and shoots dry out, the less likely the disease is to spread. Chemical sprays will also be able to reach their target more readily if the leaf canopy is open. Grape varieties and clones with smaller leaves, less vigorous growth, an open growth habit, looser bunches, and some inherent resistance to fungal issues will be more suited for growing in regions where fungal diseases, including Botrytis, are less common. These cultivars will also yield greater rewards. Some Pinot Noir growers in Germany (and possibly other countries, though I haven't seen it) split bunches in half around véraison, a technique known as traubenteilen.

This method reduces the bunch's size in half and makes the shoulders of the bunch droop, loosening the bunch and facilitating easier access for sprays, light, and air. It is difficult, if not impossible, to breed out botrytis, even with the best of intentions on the part of vine breeders. The issue lies in the fact that the disease feeds on dead and decomposing plant matter, which is something that all plants experience as their lives draw to an end. Bunch architecture, on the other hand, refers to clones with looser bunches, larger stalks, and more separated berries, which can be bred into different varieties. For example, there are a number of loose-berried clones of Pinot Noir, a fairly susceptible variety to Botrytis. These are referred to as grappes lache clones in France and lockerbeerige klone in German, of which the Mariafeld varieties are an excellent example. For some good information on these, check out the vine nursery.

Noble Rot, Pourriture Noble, and Edelfäule are examples of botrytis that affects grapes differently once they approach a potential alcohol level of about 7%. As it feeds on the berry's skin, the fungus will pierce it without rupturing it, allowing water to seep out. Each grape in the bunch will gradually shrink as a result of this slow, controlled process, which could take up to six weeks. The remaining grape juice will have a significantly higher sugar content but a much smaller volume. A few conditions must be met for this to occur: the crop needs to be cleaned up until the sugar level reaches the proper level; the weather needs to be warm, even hot; and humidity is needed for the Botrytis spores to develop. It's commonly believed that early morning mists, like those found in Sauternes and Tokaji, are the key to successful Botrytis development.

The most suitable grape varieties for this have been found to be: Sémillon and Sauvignon Blanc (Sauternes, Barsac, Montbazillac, Australia), Riesling (German Trockenbeerenauslese, Beerenauslese, Auslese), Furmint and Hárslevlú (Tokaji), Chenin Blanc (Loire sweet wines),

Gewürztraminer, Pinot Gris and Riesling (Alsace's Vendange Tardive and Selection des Grains Nobles76. In Austria, Ruster Ausbruch wines from the region around the town of Rust (which borders the lake called the Neusiedlersee), are made from various Botrytis-infected grape varieties – Furmint, Welschriesling, Pinot Blanc, Chardonnay and Pinot Gris being the main ones. The best sweet wines are those that have enough natural acidity to counterbalance the high levels of residual sugar. All of these varieties have relatively high acid levels. In some warm, arid places (like Australia and California), where Botrytis is not a naturally occurring occurrence, grapes are sprayed with a liquid infused with active Botrytis spores in an attempt to induce the disease. The fungus that causes botrytis produces an antibiotic called botryticine, which prevents yeast growth in the process of making sweet wines from infected grapes. This is likely one of the reasons these wines are sweet.

(a) slowly ferment and

(b) despite having a significant amount of residual sugar left in them, rarely re-ferment.

Peronospora, Mildiou77 Downy Mildew (Plasmopara viticola) is a wine region-wide fungus that originated in North America and was first observed in Europe in 1878. Though not all vineyards worldwide have it, it is now endemic and found in most of them. Growers in Oregon claim not to have seen it at all. It overwinters on fallen leaves and, once established in a vineyard, cannot be completely removed, though it can be managed with consistent spraying. It prefers warm, humid weather and, once the weather warms up in the spring, it will attack any green part of the vine. If left unchecked, it will also destroy flowers and berries, which will result in a total loss of crop. If impacted, grapes will shrivel and turn leathery, hence the German term for the disease, lederbeeren, which means leatherberry. If there are enough contaminated grapes in the harvest, the wine will have a distinct moldy taint.

A common pattern in the disease's spread is that the grower gains chemical control early in the growing season, but late-season outbreaks happen when the leaf wall thickens and becomes more dense, making it harder for sprays to penetrate and allow air to circulate within the canopy. The ripening process slows down in these situations because photosynthesis is decreased and leaves sustain significant damage. Although the standard physical methods of controlling downy mildew, such as opening the leaf wall to light, air, and drying winds, are helpful, spraying will still be required to completely eradicate the disease. The discovery (by accident) in 1885 that copper-containing sprays were effective against Downy mildew led to the creation of Bordeaux Mixture, a highly effective protective spray against the disease that is a mixture of slaked (or hydrated) lime and copper sulphate. Its distinctive dusty blue/mauve traces are visible in vineyards all over the world. It is extremely rainfast and can help control the disease during the dormant season if sprayed after harvest but before winter. Bordeaux Mixture's drawbacks include the need for fresh preparation each time it is used, its instability, and its short shelf life.

Furthermore, it is merely a protector—not an eradicant. The main problem with its use, though, is that it raises soil copper levels, which are poisonous to earthworms and other soil animals. Over the past few decades, the amount of copper that can be used in EU vineyards has gradually decreased, and as of right now, the maximum allowed is 4 kg of copper per year, averaged over seven years. Bordeaux Mixture is used far less frequently now that there are alternative, easier-to-store products available. The most popular ones contain copper hydroxide and copper oxychloride. They can be stored, don't need to be prepared, and are less likely to linger in the soil. Fortunately, there are numerous fungicides that do not contain copper that can be used to control Downy mildew. These fungicides were developed to treat Potato Blight, a disease that is very similar to Downy mildew.

Producers who practice organic and biodynamic farming face in relation to Downy mildew. Growing grapes organically and Biodynamically becomes more challenging as the amounts of copper permitted are constantly lowered (it is eventually assumed to a point where the amount allowed is no longer effective). There are no common varieties that are resistant to natural protection, despite the fact that some PIWI and older hybrid varieties do. How this challenge is handled in the future is still to be seen. Oidium tuckerii, also known as Uncinula necator, is a fungal disease that is indigenous to North America and causes powdery mildew. Although it was first reported there in 1834, it did not significantly harm local varieties and did not generate much excitement. It was only at that point that vine growers began applying sulfur to their vines—sprayed in colder climates and dusted in warmer ones—which, when applied promptly, is an incredibly inexpensive and efficient remedy.

Additionally, it was discovered that sulfur contributed to higher yields and accelerated the harvest by 7–10 days. These two effects accelerated the adoption of sulfur as a useful remedy. Sulfur's protection against Oidium was just one of many advantages; it also has some effect against other illnesses and mites. Applying sulfur either right before or during flowering is not advised if the temperature is predicted to rise above 30°C because scorching of the flowers could result in crop loss. If the weather is conducive to the disease, sulfur spraying should continue all summer long.

Unlike most fungal diseases, oidium does not require water to spread from one part of a plant to another, making it one of the few that are worse in dry years than in wet ones. Sufficiently humid air is sufficient. It thrives inside shaded canopies and prefers warm, humid weather (20°C to 27°C) and still, dry canopies. When the conditions are right, it overwinters on old wood and, frequently early in the growing season, begins to produce spores that are dispersed by the wind and attack any part of the vine. If early infections are permitted to occur, sulfur cannot be used for prevention; instead, more costly chemicals will need to be used to completely eradicate the infection. The young berries split open to reveal the seeds when Oidium attacks them, covering them in a distinctive white powder. Damaged berries typically never ripen because Oidium attacks them early in the growing season. The juice is only tainted if a significant number of damaged berries are still on the vine when the fruit is (machine) harvested. Depending on the precise formulation, sulfur's harvest interval ranged from 21 to 56 days.

However, approximately 20 years ago, sulfur's legal classification shifted from pesticide to fertilizer. This was due to the fact that sulfur levels in the atmosphere and, consequently, soil have significantly decreased since the use of coal and coal products for energy production and heating in both domestic and industrial settings has decreased. Once unheard of fifty years ago, many arable farmers now discover that their soils require sulfur addition. Growers are therefore free to use as much and as often as they choose. Some nations do not have harvest intervals for sulfur, but sulfur should be sprayed or dusting done well in advance of harvest because if residues are left on the grapes, fermentation may produce hydrogen sulfide (H_2S), which has a rotten egg odor. Fortunately, Oidium is one of the most predictable diseases, making preventative measures easy to implement. It is fairly easy to predict when Oidium will be releasing spores and, consequently, when spraying will be most effective by combining temperature, humidity, and rainfall readings. Because oidium is a disease that does not manifest visually until well into its second generation—that is, when it is too late to take preventative action—this forecasting is essential. Light mineral oil is being used by some conventional and organic farmers in place of sulfur. This oil kills the fungus's cells and also works similarly on certain mites and spiders, including Botrytis. In addition to sulfur, a variety of other substances and mixtures are employed to manage oidium, and in recent times, strobilurin-based sprays have shown to be highly successful. Potassium bicarbonate sprays have also been demonstrated to be successful.

Other Diseases

Anthracnose

This is also known as Bird's Eye Rot and Black Spot, is a disease originating in Europe that significantly reduced crop yields prior to the introduction of Powdery and Downy Mildew. It usually manifests itself quite early in the season and is a disease of damper climates. Nonetheless, anthracnose is rarely a standalone issue and is readily managed with Bordeaux Mixture and other fungicides containing copper.

Armillaria Root Rot

This is also known as Honey Fungus, Oak-Root Fungus. This fungus, which is often found in woodlands and forests, can damage vine roots, particularly in areas where vineyards have been planted on formerly forested land. The plant will eventually wilt and show signs of diminished vigor. Vines often die in a matter of weeks, and this can happen quite suddenly. When planting on previously forested land, it is important to be aware of the possibility of disease symptoms on the roots of the trees that are currently growing there.

Unfortunately, control is nearly impossible to achieve outside of extremely small areas that can be subjected to soil fumigation, typically using methyl bromide. This is not a practical solution for large areas. When this fungus is known to be present, the most practical course of action is to leave the land fallow for as long as possible, deep-plowing it periodically to expose and kill old tree roots. There is currently no developed resistant rootstock, despite research efforts.

Bacterial Blight

Although not common, bacterial blight is a problem in many parts of the world and when it does occur, it can lead to vineyard abandonment because there is no known treatment. Infected vines eventually become unprofitable as their yields decline and they gradually lose their vigor. Pruners must sanitize their secateurs in between vines as the only means of controlling the disease. It is more severe in areas with summer rainfall than in drier regions, although copper sprays can help with control. Black Rot Guignardia bidwellii is the organism that causes black rot, also referred to as Le Black-Rot in France. Once it is in a vineyard, it is impossible to remove. North American in origin, it was brought to French vineyards in the 1880s. Maintaining an open and airy canopy, getting rid of damaged material (especially unpicked or fallen grapes), and maintaining a clean vineyard all help with control. For the majority of vineyards, chemical control will be required, though. As soon as the temperature and moisture levels are ideal, black rot will begin early in the growing season. Spraying needs to be done continuously until well after flowering. Fortunately, it is managed by the same range of drugs that manage the other major fungal diseases—Downy Mildew in particular—so in actuality, it is not a condition that needs to be treated exclusively.

Crown Gall

Agrobacterium vitis is the cause of Crown Gall, also known as Black Knot, and it has been identified in European vineyards since the mid-1850s. It is feasible for entire batches of vines to become infected because grafting is the most popular method of distribution. The most frequent times that vines get infected are when an established vine sustains damage from a severe winter frost that splits the trunk, or when the vine sustains mechanical damage from a hoe, weeder, or mower. The bacterium, which is frequently found in vineyard soils, then settles down and produces a big, woody gall that can grow up to the size of a golf ball, sometimes even bigger. As a result, the graft is frequently destroyed and the vine eventually dies. Winter injury to the graft can be avoided by plowing close to the vine so that soil is piled up around and over the graft before the winter (buttage). The earth needs to be hoed and ploughed away from the graft in the spring. In German, the illness is referred to as Krebs (Cancer) or Mauke. Growers in vineyards in New York State train several trunks up from the base so that in the event of an infection, one can be severed and the vine will still bear fruit. It is occasionally regarded as one of the many diseases of the trunk.

Flavescence Dorée in Grapevine Yellows

The word "grapevine yellows" refers to a collection of related illnesses brought on by phytoplasma, which are microscopic (or sometimes even submicroscopically tiny) bacteria-like organisms that enter the vine's sap. Flavescence Dorée was first identified in the Armagnac region in the 1940s, and since then, it has spread throughout Europe and is currently found in many wine regions. It is primarily transmitted by two vectors: the use of infected material during the vine-grafting process, and sap-feeding insects (leafhoppers) that can jump from one vine to another. The leaves of infected vines will curl, turning yellow as they ripen and quickly affecting yields. The disease has no known chemical cure, so the only methods left to combat it are limiting the number of plants (weeds and grasses) that the vectors can infect and feed on, managing the vector population, and improving sanitation in vine nurseries. It was initially observed in Australia in the middle of the 1970s, has spread to various wine regions, and causes significant crop loss. In France, areas with Flavescence Dorée require new vines to be heat-treated before they are allowed to leave the nursery. This will lessen the occurrence of grapevine yellows and other illnesses.

New vines can also be treated by immersing them in hot water at 50°C for 30 minutes prior to planting. Pierce's Disease, also known as Anaheim Disease or PD, was initially discovered in vineyards in Anaheim, California, in 1892. Since then, it has spread throughout North, South, and Central America. Though not inconceivable, its occurrence in other regions of the world is less common. Xylella fastidiosa, a bacterium that inhabits the plant's xylem and is dispersed by a variety of insects, is the source of Pierce's disease. The disease in California is controlled by limiting the spread of two leaf-hopper species that are accountable for the bacterium's spread: the glassy-winged sharpshooter and the blue-green sharpshooter. The first signs include unusual leaf staining and marking, which are typically noticed at the end of the first summer of infection. The vine is less vigorous in the second year of infection, with stunted shoots bearing few viable fruit buds, which results in a significant loss of crop. Younger vines are more vulnerable than older ones, and an infection will typically make a vineyard unprofitable. Since there is currently no known chemical treatment for Pierce's disease, prevention efforts center on lowering the number of sharpshooters in the area around the vineyard. The spread of the disease was primarily caused by the blue-green sharpshooter when it initially showed up in northern California. Since this insect liked to spend its summers in watercourses, it was beneficial to keep these weed-free and plant vines a distance away from them. The glassy-winged sharpshooter, on the other hand, is much harder to control because it is a more omnipresent insect that is not restricted to riverbanks.

Although the susceptibility of the various varieties varies, California could suffer greatly from the disease because Chardonnay and Pinot Noir are easily damaged. It was found on olive trees in Puglia, southern Italy, in 2013, and it has been gradually moving north, affecting oleander and olive trees alike. Although it hasn't yet had an impact on vines in Europe, it most certainly will. With winter temperatures rising due to global warming, sharpshooters may be able to overwinter and contribute to the disease's spread throughout Europe. Grapevine viruses: Since grafted vine planting became commonplace, viruses have proliferated across the viticultural industry. Nurseries that procure plant material from virus-infected sources, including scions and rootstocks, have been the primary means of dispersing these economically devastating illnesses. Once within a vineyard, viruses are frequently dispersed using more organic means, with nematodes and insects acting as vectors to move the infection from one plant to another. Viruses are a constant risk in certain regions of the world, particularly in nurseries with poor sanitation, few government regulations, and ungrafted rooted cuttings that growers are free to plant. A number of viruses actually only really affect people in the nursery industry because they either cause graft failure or vine death during rooting before sale, which is not good for business. The following are the most significant factors for grape growers.

Bark Corky

One of the most prevalent viruses is corky bark. When a vine becomes infected, its leaves turn red or reddish-yellow and curl downward, refusing to fall off during dormancy or even for a brief time after a frost. Canes will also develop grooves, and parts of the main stems will enlarge and take on a corky appearance. Not all species and varieties are affected equally, and some—particularly hybrids and native North American vines—may sustain far more damage than Viniferas. Like most viruses, there is no known cure; the only effective treatment is to destroy and remove the infected vines. One of the oldest viruses, Fanleaf Degeneration - Fanleaf Virus, Court-noué Fanleaf has been a part of European vineyards for more than 200 years. As the name implies, the virus causes leaves to grow and distort into the shape of a fan. It can also cause the leaves to turn a mottled green color. Furthermore, growing tips may divide into two or even three distinct shoots, each with canes and nodes that are flatly distorted. The disease's second and third phases are characterized by yellowing of the leaves in patches or veinbanding (yellow mosaic and veinbanding), which eventually causes yields to decline and eventually cease completely. The only treatment for this virus, like most others, is to remove and destroy the contaminated material, let the area lie fallow for as long as possible (the virus can linger in old roots for up to six years), and then replace it with virus-free plants. Since nematodes carry the Fanleaf Virus, planting on rootstocks resistant to nematodes will help control the disease.

Rolling leaves

Of all the viruses, leafroll is another one that is very common and probably does the most harm to the economy. Although leafroll is actually caused by multiple distinct viruses, they are often referred to as one because of how similar their symptoms are. Nurseries that produce grafted vines must take extra precautions because the disease is difficult to control because symptoms do not always appear on diseased vines, especially on American rootstock varieties. As the name implies, the symptoms include the leaves rolling and showing faint veining, along with a color shift, particularly in the fall, from green to bronze to red. In actuality, many images of vineyards displaying "glorious autumn color" are excellent illustrations of vines that are severely infected with the leafroll virus! Vine disease will quickly lower yields, cause grapes to ripen more slowly due to malic acid, and prevent grapes from reaching the necessary sugar levels. In addition to the grafting process, rare instances of its spread in South Africa and New Zealand have been linked to vectors like the Mealy Bug. Due to the spread of infected material during the apartheid era, when plant material from abroad was (officially) prohibited and vines were produced by a very small number of not very independent (and not very careful) nurseries, it is regrettably quite common in South Africa. Now that infected vineyards are being removed and replaced with plants free of viruses, this is beginning to change.

Nepoviruses

A family of 13 viruses known as nematodes spreads the nemoviruses. These comprise tobacco, tomato, and Arabis mosaic (nepo)virus; they are closely linked to fanleaf degeneration. The incidence of these viruses can also be decreased in nurseries by applying heat treatment to cuttings and immersing grafted vines in hot water. They can survive in infected plant material for up to ten years, and once they are in vineyards, they are hard to eradicate. Under a microscope, neoviruses all have a similar polyhedral structure.

Red Blotch

This is damaging vineyards across North America; it was only identified as a distinct virus in California in 2012. As its name implies, it first manifests as red patches on leaves before moving on to the grapes. Here, it causes the berry to shrivel, change color, and postpone ripening, all of which have a substantial impact on the crop's quality and economic value. The three-cornered alfalfa tree hopper, which is its vector and causes vine-to-vine transmission, is too widespread to be managed with pesticides. Right now, the only way to combat Red Blotch is to grub and replant with virus-free vines after the vineyard reaches an unprofitable stage.

Rugose Wood

Rugose Wood, Complex Rupestris stem-pitting, Kober stem-grooving and Corky bark, stem-grooving are just a few of the viruses that can cause vines to become rugose wood. Similar to other viruses, vine nurseries are more concerned about this than grape growers are because of losses from bad grafts and deaths of young vines while they are in nursery beds. Also known in Italy as Legno Riccio. Trunk Diseases: Though many of them have existed for decades, some for over 2,000 years, trunk diseases (TDs) are a collection of illnesses and vine maladies that have become more common and harmful to vines in recent decades. But in recent times, they have also garnered more attention, been the subject of more conversations, and gained some notoriety in certain areas and varieties. "Young Vine Decline" is a general term used to characterize issues that have impacted numerous regions with a wide range of varieties and are brought on by multiple organisms. These harm the woody portions of the vine and frequently contaminate the young grafts, leading to graft failure and eventual plant death. The organisms that cause TDs are frequently found in scion mother-gardens and rootstock plantations, and as a result, they are present at every stage of a vine's life, including pre-grafting, grafting, planting, and establishment. Pruning cuts in the vineyard can spread them from vine to vine. They propagate from one side of a vineyard to the other over time and are also wind-blown from vine to vine.

The causes of the rise in TDs are far from clear-cut, and the ones that have been suggested are frequently contested by various viticultural groups. Do nurseries take adequate measures to prevent the spread of TDs? Is it because vines are more likely to get TDs because they are less susceptible to viruses now? Is there a global warming? Is drought putting more stress on vines? However, one thing is for sure: in recent decades, new diseases have infected a large number of hardwood plants. Dutch elm disease, sudden oak death, red band needle blight, horse chestnut bleeding canker, and most recently, ash dieback, have all been observed in the UK alone.

Thus, it appears that vines are not the only ones who observe and experience new diseases. It is important to emphasize that the majority of vines, like the majority of humans, contain both good and bad bacteria, as well as both types of gut microflora. Our bodies are able to handle conflicting factors when we live in a normal environment with enough food and drink, no physical or mental stress, and adequate rest and sleep. However, when we skip meals, become dehydrated, sleep too little, or both, harmful bacteria and microflora begin to proliferate and we become sick. Vines are remarkably similar.

Any new vine can be tested for TD organisms, and you will typically find a wide variety of them. Some new vines, such as those with Petri Disease and Bot Canker, will exhibit the classic staining of TDs when cut in half. However, in each of these situations, the ostensibly "diseased" vines frequently continue to grow and prosper. But this isn't always the case. Generally speaking, when vines are growing in a stressful environment, they won't flourish and will eventually fail.

To try to decrease the impact of TDs, there are a few basic hygiene guidelines.

Avoid pruning when it's wet.

Aim to prune as late in the spring as possible when the weather warms.

Apply wound paint to all large pruning wounds.

Maintain vineyards neatly by getting rid of as much pruning wood as you can. Retain away from training and pruning schemes involving extensive old-growth cordons.

Here are a few of the most typical vine TDs that can be found. Cylindrocarpon, or "Black Foot," is a fungal disease that causes a reduction in vine vigor, resulting in stunted shoots, short internodes, and unproductive vines.

Like nearly all TDs, Botryosphaeria Canker, also known as Black Dead Arm, affects both young and old vines and eventually results in vine death. When cutting through cordons or trunks, a distinctive dark stain—typically in the form of a V—is revealed. Cutting back to just above the graft and choosing a handy "water-shoot" to create a new trunk and/or cordons can sometimes revitalize a failing vine. Esca, also known as Black Measles or Apoplexy, is more common in warm to hot climates and is caused by one of several different fungal pathogens. Usually, the first signs of damage appear as light-colored patches between the veins, which eventually cause the leaf edges to turn brown and become necrotic.

Dark speckling, also known as "black measles," will first appear on the grapes before they shrivel and drop. The vine will suddenly wilt and die in severe conditions. Eliminating large pruning wounds, avoiding pruning systems with a lot of permanent wood (long cordons, GDC, Sylvoz, etc.), and cleaning the vineyard of old wood all contribute to control. There is also "Apoplexy," or the sudden death of vines, known to the French as folletage or tylosis, and "Young Esca," or black xylem decline, which is brought on by Phaeoacremonium and other fungi. In the world of winegrowing, Eutypa - Eutypa Dieback, Eutypiose Eutypa is widely distributed, particularly in areas with more than 600 mm of annual rainfall. It doesn't care about the site or variety. Grafting is a common way for the fungus Eutypa lata, which causes it, to enter vineyards. But it rarely appears in young vines; instead, it prefers to wait until the vines reach about 10 years of age, at which point their vigor begins to decline. The first noticeable symptoms typically appear when spring shoots are stunted and fail to thrive, leaving them with short internodes and malformed, chlorotic (yellowing) leaves.

Because of this damage, which is frequently limited to just one arm (of a spur-pruned vine), the disease is sometimes misidentified as dead-arm, a term more commonly used to describe phomopsis. Cutting through the arm or trunk of a suspected infected vine is one method of verifying the disease's existence. A wedge-shaped area of the arm or trunk that is dead and frequently makes up about one-third of the total area of the arm or trunk is indicative of an infected vine. This area will grow until the vine dies as it becomes more and more infected.

In actuality, there is no quick fix or long-term treatment for eutypa. Large pruning wounds can be painted with a fungicide-containing paint or cleaned if Eutypa is known to be present in the vineyard. With some adaptation, secateurs can now spray fungicide as soon as a cut is made, though this definitely slows down the pruning process. Applying fungicide paint to every wound is also an expensive and time-consuming process, but both of these measures slow down the disease's progression. Later in the season pruning will also be beneficial because, given enough warmth, pruning wounds heal more quickly.

Additionally, the vine's natural spring bleeding will provide some protection by sealing the cut surface. Given that Eutypa favors bodies of old wood, it will be beneficial to steer clear of training systems that contain a significant amount of old wood, like spur-pruned cordons, and instead use cane-pruned systems. Retraining the trunk completely has helped revive some Eutypa-infected vines by taking a water-shoot that emerges from near the ground (but above the graft). This procedure extends the vine's productive life by a few years, even though it may cause the vine to lose its crop for up to two years while it grows new fruiting wood.

Although, Eutypa is also found in several non-Vitis species, removing old wood from the vineyard after pruning and burning it will help stop its spread. However, this is not always effective. Petri Disease: Black Goo: This disease, which originates in the rootstock and scion wood, is related to Bot Canker and Black Foot. Petri disease-affected vines, when cut in half, release tiny, pin-prick-sized black, oily particles; this is how the plant gets its nickname, "Black Goo." Another common viticultural disease is Phomopsis, also known as Phomopsis Cane and Leaf Spot, Dead-Arm, Excoriose Phomopsis (Phomopsis viticola). This disease is more common in areas that receive consistent summer rainfall. Similar to Black Rot, it is managed by chemicals used to combat Downy Mildew; additional treatments are typically not required. Spraying may be necessary for recently planted vines because it can be an issue in nurseries.

CHAPTER SIX

PESTS THAT AFFECT GRAPEVINES

In addition to physically harming leaves and impairing their ability to photosynthesize, pests can also live on leaves, weakening the vine overall by sucking their sap. Lastly, pests can attack roots, reducing the vine's ability to absorb moisture and nutrients. Finally, pests can act as vectors, feeding on the sap of vines and spreading diseases, viruses, and other ailments from one vine to another. Certain insects need to be managed because they could physically harm the fruit. It might not be worthwhile to spray for the bug if the damage is essentially only cosmetic or, in the case of wine grapes (as opposed to table grapes), not economically significant. Insecticides and acaricides are used to kill insects and mites, respectively, but most growers use very little of these products because most naturally occurring predators will take care of the job for free.

Pesticide use may be nonexistent in many areas, particularly in those where vineyards are somewhat isolated from one another and natural predator populations are high. Sulfur (as well as a few other chemicals) sprayed to prevent powdery mildew (Oidium) will also prevent light infestations of mites and other pest species. In order to determine whether populations of harmful insects are large enough to justify spraying, sticky-floor traps coated with the female pheromones of the target insect are frequently used for this purpose. Males are drawn to these traps, where they become stuck to the floor and can be counted.

Frequently, the number proves to be many times higher than the number captured on sticky floor traps. A book many times larger than this one would be needed to list and describe all the pests that can harm vines, because there are so many of them. Numerous pests are confined to particular areas or regions with particular climates, and they might not even exist in other places. As a result, the ones that have been described in detail below are the most common and destructive.

Beetles

Many different types of Beetle's prey on vines. In Europe, two beetles that cause enough damage to warrant regular use of insecticides are the êcrivain (Bromius obscurus) and cigarier (Rhynchites betuleti). Additionally, some beetle species—referred to as borers—have larvae that eat away at the woody portions of vines, such as the cordons and trunks of spur-pruned vines. Vegetables have been known to sustain harm from borer damage in locations as distant as California, Oregon, and the Hunter Valley.

Borers are generally quite confined. The larvae of moths known as cutworms reside in the ground or beneath the bark of vines, emerging at night to feed on the leaves of the plant. They can attack newly planted vines and slow down their establishment, though they are not really a commercial problem on established vines.

Erinose Mite

Grape Erineum Mite, Grape Leaf Blister Mite: Colomerus vitas, the only known host of the Grape Erineum Mite, causes little harm to the grapevine. However, because of its resemblance to the leaf form of Phylloxera, viticulturists are frequently taken aback when they see it. This mite causes very unsightly swellings on leaves, usually giving them a reddish tinge. Fortunately, sulfur usually suppresses the mite. Drosophila, commonly referred to as the Vinegar Fly or Pomace Fly, is a small insect that does not pose a significant threat to human health. But it will visit fruit that has already been damaged (possibly by hail, wasps, or other insects), particularly in the later stages of ripening when sugar levels are high, and it will spread acetic acid bacteria, which causes "sour rot" and increases the volatile acidity (VA) level in wines made from this fruit.

Similar harm is caused by the Mediterranean fruit-fly, which is an issue in warm to hot climates (South Africa and Australia) and is not a Drosophila species. Vineyards that are near early-ripening fruit crops, like figs, peaches, and apricots, might sustain more damage than vineyards that are isolated or in areas where grapes are the only crop. Also see the Spotted

Wing Drosophila entry. Grasshoppers, Locusts: As members of the same family, grasshoppers and locusts can attack vines, consuming their leaves and inhibiting the vine's capacity to ripen. Spraying is the standard procedure when the numbers get dangerously high. Grasshoppers frequently move from grassland or other crops to vines, particularly when the latter are well-irrigated and have an abundance of green foliage while the former are suffering from drought. Harmonia axyridis, commonly known as the Multicolored Asian Lady Beetle or Harlequin Ladybird, is a native Asian species that was brought to the United States to aid in the biological control of other pests. Though previously thought to be too cold for it, it has now made its way to Europe, where it has already caused spoiled wines. It stains grapes and wine made from them with a substance called "reflex blood," which it exudes from its legs when disturbed and has an unpleasant taste and smell.

Leafhoppers

In addition to feeding on and killing leaves, leafhoppers also act as carriers of more serious illnesses like Pierce's disease, Flavescence Dorée, and certain viruses, causing harm to the vine directly. The two main methods of controlling leafhoppers are to spray insecticides and destroy their habitat, but these methods are not 100% effective. A small wasp known as Anagrus epos is a natural predator that feeds on the larvae of leafhoppers; however, it is vulnerable to sulfur sprays. The term "leaf-rollers" refers to an insect species whose caterpillars cause damage to a vine's leaves, causing them to roll up and reduce the amount of leaf area available for photosynthesis. This insect species is unrelated to the Leafroll Virus. Several leaf-rolling species are restricted to specific vineyard areas and are primarily managed through the use of selective insecticides. Natural predators may be effective against low levels of attack in some areas. When the fruit is harmed by caterpillars, it could make room for other organisms like Botrytis to do even more harm.

Magarodes

Ground Pearls Phylloxera-like soil-dwelling insects that defoliate vines are called Magarodes or Ground Pearls. They cause the vine to become stunted, lose its energy, and eventually die. In addition, margarodes smell bad and don't seem to respond to most kinds of control. No rootstocks have been created to withstand them. Fortunately, it seems that they are limited to South African vineyards, though they have occasionally been spotted in other nations (Chile).

Mealy Bug

Mealy bugs come in a variety of species, some of which are harmful to vines. Though they genuinely cause minimal mechanical harm, they can coat clusters of grapes in a substance called honeydew, which attracts a black sooty mold.

If this mold is brought into the winery, the grape-based wine will become tainted. Certain mealy bug species, particularly Leafroll, have also been connected to the spread of viruses. Many ant species are drawn to the honeydew and will attack natural predators of mealy bugs and other insect pests, even though they won't harm vines.

Mealy bug control is a laborious and time-consuming process that necessitates meticulous observation of insect activity and abundance before applying insecticides in an efficient manner. Certain mealy bug species are notoriously hard to manage because their eggs cannot be sprayed because they overwinter beneath the vine's bark. In greenhouses, removing the bark from vine trunks and cordons and applying a fungicide to them is a yearly task to prevent mealy bug infestation.

Mites

Mites are microscopic insects that resemble spiders and live on the leaves of vines. Certain species, like the grapeleaf rust mite, Pacific spider mite, red spider mite, and yellow mite, all feed on the leaves of plants, decreasing the plant's capacity for photosynthetic energy and delaying the process of ripening. Vine leaves that are severely infested will almost entirely turn red, giving the impression that they are naturally colored for fall. Fortunately, the sulfur used to control powdery mildew slows down mite activity; however, if attacks become severe, an acaricide will be sprayed. Predatory mites are another tool growers can use to manage harmful ones; however, sulfur sprays do not usually distinguish between beneficial and harmful mites.

Moths

Quite a few moths harm vines; however, it would be more accurate to say that their larvae do the damage, as they burrow into the young grapes upon hatching. While using pesticides to repel moths is a common practice, pheromones are increasingly widely used in many nations to induce "sexual confusion" (confusion sexual) in the species' males.

The most widely used system in vineyards is one in which the pheromones are suspended on vineyard posts inside tiny, typically brown capsules. They function by tricking the male, who spends more time observing the capsules than mate-making with female moths. The only disadvantage is that all growers in the impacted area must use these capsules; otherwise, all of the males congregate on the one or more vulnerable vineyards, mate, and then all of the females with their eggs scatter over the region. When the male moths are active in the morning and evening, some systems employ "puffers" to release pheromones into the atmosphere. Pheromone and feeding traps are often hung to keep an eye on these moths' activity among the vines because their attacks are usually irregular.

Additionally, a variety of insecticides can be used to control various predator species that can be employed against moths. Even though the damage caused by the larvae may not be very severe, the grapes' wounds will be perfect for illnesses like botrytis to take hold. Although a few of these moths have been spotted in the UK on occasion, if the climate continues to warm, they may begin to make an unwanted appearance.

The most frequently observed moth species are as follows: Cochylis - Eupoecilia ambiguella (also called Traubenwickler) Eudemis - European

Grape Moth - Lobesia botrana

Eulia - Argyrotaenia pulchellana

Grape Berry Moth - Polychrosis viteana

Grape Vine Moth (NZ) - Phalaenoides glycine

Orange Tortrix Moth - Argyrotaenia citrana

Pyrale - Sparganothis pilleriana

Nematodes

Nematodes, also called eel-worms, are microscopic worms that live in soil and can feed on vine roots, depriving the vine of both water and nutrients and weakening the vine so that it becomes a prey to other diseases. One of the most common nematodes causes knot-like growths on vines: the root-knot nematode. There are other nematodes that act as carriers of viruses like Fanleaf. In actuality, nematodes are difficult to control chemically and are difficult to eradicate. The best approach is to use nematode-resistant rootstocks in conjunction with improved nursery hygiene. Before planting, rootstocks can be soaked in 50°C hot water for 30 minutes to kill any nematodes without damaging the plants.

Scale Insects and Rootstocks Phylloxera

Numerous species of scale insects exist, and they are all responsible for weakening the vine by feeding on its sap. Certain species also release honeydew, a sticky substance that attracts the same black sooty mold that is found on mealy bug-damaged vines and produces comparable effects. Insecticides are used to control scale insects.

Slugs and Snails

While slugs and snails rarely cause problems for established vineyards, they can cause significant harm in newly planted areas, particularly in areas where vines are shielded from rabbits and other predators by individual guards. Slugs and snails, which eat leaves and tender shoots, seem drawn to the warm, frequently damp environment within the guards. Generally effective, but depending on the severity of the issue, it might need to be repeated, sprinkling anti-slug and snail pellets into the guards. On mature vines, snails have occasionally been observed to cause minor damage to the leaves and shoots of the early spring growth, but not enough to require treatment.

Spotted Wing Drosophila

Drosophila suzuki, also referred to as "SWD" informally, is a fruit fly that has been causing havoc on a variety of soft fruits in Europe, including apricots, cherries, blueberries, grapes, nectarines, pears, plums, peaches, raspberries, and strawberries. The larvae of SWD, in contrast to common fruit flies, hatch inside the affected fruits, feed on the pulp, and only emerge when they are fairly well developed—something that consumers do not want to witness while they are eating the fresh fruit. Increased volatile acidity (VA), which is undesirable in wineries and could potentially reduce the value of crops grown by growers under contract, is a result of SWD's attack on wine grapes.

Thrips

Often called thunderbugs, rips are tiny, black-winged insects that cause stunting to young shoots and feed on vine pollen during flowering. There are multiple species of rips. The grapes become russeted and scarred as a result, which could lead to splitting as the grapes ripen later in the year. There isn't much evidence to support the theory that thrips cause poor fruit set. Table grapes suffer more economic harm from them than wine grapes do because the damage is primarily cosmetic rather than real. Western Grapeleaf Skeletonizer: This moth lives up to its name by turning grapevine leaves into skeletons. However, it is the larvae that cause the damage, not the adult moths. In any case, it's a pest that needs to be sprayed if it's discovered because, if ignored, it can totally defoliate a vineyard.

Rootstocks and Phylloxera

The devastator, Phylloxera vastatrix, is the primary cause of the vast majority of vines grown worldwide, although it's not the only one. Vines are grafted onto rootstocks. Despite being more widely recognized as Phylloxera, its true name is currently Daktulosphaira vitifoliae.

Phylloxera Grape

Phylloxera inhabits the wild vines known as native American vines, which are found in the eastern and southern regions of North America. Through co-evolution with its host, it was able to establish a symbiotic relationship with these wild vines, feeding on their leaves and roots and occasionally weakening but never killing them.

Many species of wild vine have evolved over millennia of coexisting, and their roots are resilient enough to withstand insect damage. The primary defense mechanism of the American vine is its ability to heal the wounds left by the insect as it feeds, protecting the vine from invasion by other pathogenic organisms like bacteria and fungi, which are the primary cause of damage to Vitis vinifera vines. Furthermore, the sap found in the roots of native American vines makes the insects less inclined to stay on any one piece of root for an extended period of time. Simply put, the sap clogs the insect's feeding apparatus, causing it to look for a better food source. This disrupts and slows down the insect's multiplication.

Additionally, there is evidence that suggests something in the roots is responsible for juvenile deaths or at least stunted growth. In summary, the American vines' roots provide an uneasy habitat for the insect, both during feeding and during breeding. Since the roots of these plants are uninhabitable, the majority of the louse population on American vines reproduces on the leaves during the summer. Thus, there is little activity on the roots. This behavior contrasts with that of other vine species, particularly viniferas, where it favors the roots over the leaves. It can lay its eggs and breed here, below ground, in a relatively safe environment. It feeds on the roots, weakening them by removing photosynthates (starches and sugars), which permits harmful bacteria and fungi to further attack the root structure. The vine eventually wilts and slowly dies, usually during a stressful period that may be related to the crop or the climate. Even though phylloxera is a very slow-moving insect—it took decades for it to spread throughout Europe—all or nearly all vinifera vines will eventually suffer damage from its own roots. It is hard to do grape phylloxera justice in the limited space allotted in these pages, but it must rank among the most fascinating living things on the planet. Its life cycle is full of twists and turns, and it can alter itself depending on the conditions of its host and surroundings. It is a difficult pest to deal with, let alone defeat, due to its adaptability to its environment, distinct behavior in sandy, clay, or chalk-rich soils, dislike of extremely hot weather and wet soils, and capacity to distinguish between pure American vinifera and hybrids. Its adaptation to its host species and surroundings, however, not only made it a terrible issue but also provided the key to its defeat.

Phylloxera, a tiny yellow aphid measuring only 1 mm in length, is a parthenogenetic female. This means that she can lay eggs without a male's attention and, presumably as a result, nearly all of the eggs she lays hatch into females. A single female can give birth to up to seven generations in a single summer, with the first-generation giving birth to up to six more, the second to up to five more, and so on, as if this wasn't enough to ensure the species' survival. The weight of aphids would be ten times that of a typical grape crop if all the eggs hatched and no progeny were lost due to the enormous multiplication factor!

This egg-laying occurs mostly on the leaves of native American vines. The leaves of an infected vine will be covered in raised lumps that resemble warts and range in color from reddish-brown to greenish-white, depending on the host species. These are called galls, and they resemble the ones that are occasionally observed on the leaves of oak trees, which are the host trees of Phylloxera quercus. The unhatched eggs in these galls eventually hatch, and the resulting aphids spend the summer feeding and reproducing on the nearby leaves until the leaves begin to fall, bringing with them the Phylloxera's food supply. The aphid then makes the decision to spend the winter underground, and some of the summer adults transform into crawlers and make their way through soil fissures and down the trunk to the vine's closest roots to the surface. Here, they continue to feed and reproduce unhindered by the elements or by humans, birds, or other insects. They also inject their saliva into the roots of the vine, resulting in what are known as root galls.

Similar to leaf galls, these galls contain the eggs that have not yet hatched. Life moves at a slightly slower pace than it does on leaves because it is colder below ground and harder to move through the soil. It doesn't make their presence there any less likely or dangerous, though. Large colonies of insects will grow from a single below-ground adult, moving from vine to vine and from root to root. Fertile adults can move a short distance on their own; they will come up from beneath the surface and scuttle across the ground. Some will be carried by humans and can spread from vineyard to vineyard and even region to region by being fastened to boots and shoes, tractor tires, harvesters, and picking bins. There are species with wings that can fly a short distance (roughly 100 meters) and lay eggs that hatch into both male and female progeny in some warm climates.

Eventually, the female offspring of this winged form lays what is known as the winter egg, which hatches into a female that is known as the fundatrix, who becomes the mother of a whole new generation in a new location and is akin to a queen bee or wasp. A number of colonies in North America, including Virginia and Carolina, were established with the goal of planting vineyards. When the early settlers discovered how many native grapes there were, they made a vain attempt to turn them into good wine. Native American Indians seemed to enjoy the wine made from these native varieties, but it was found to be inedible, at least to the palates of Europeans.

Next, attempts were made with Vinifera vines, which were undoubtedly brought in with great anticipation and zeal. Though many attempted, none were successful. Thomas Jefferson invited renowned Italian grower Filippo Mazzei to plant European varietals in a vineyard on 162 ha (400 acres) of land next to his estate at Monticello, Virginia, in 1773. The vines were either killed or damaged by Oidium (Powdery Mildew) or Phylloxera, even after they were planted and replanted over. In the years that followed, Jefferson and others found that only native varieties would bear fruit, and a business centered around vines from species like V. labrusca and V. rotundifolia developed. Growing widely, Scuppernong is a commercially successful rotundifolia variety with an unlikely name that is still grown in some states.

One kind description of its flavor is "musky." Plant breeders quickly gained proficiency in creating varieties with strong resistance to Oidium and restricted tolerance to Phylloxera. Natural hybrids between American varieties and viniferas were also generated. These breakthroughs would come in very handy when Old World viticulturists arrived seeking a Phylloxera remedy.

CHAPTER SEVEN

MINOR AND TRACE ELEMENTS

Boron (B)

Boron is a trace element, but deficiencies in it can significantly impact vine health, particularly in regions with high rainfall and sandy, acidic soils. Poor fruit set resulting from inadequate pollen tube growth is the most obvious effect, which is followed by unevenly sized grapes with an excess of small berries. Additionally, shoot growth is frequently distorted and irregular, and yields are frequently negatively impacted. When irrigation water has high concentrations of the element, excess boron can be an issue, so levels need to be closely watched. Since boron does not move around in plants very well, foliar feeding is the best course of action when a vine exhibits boron deficiency symptoms.

Calcium (Ca)

Calcium helps shield grapes from microbial attack and is necessary for healthy vine growth and fruiting. A lack of calcium is never an issue in many vineyard soils because these soils are naturally high in the mineral. Low pH (below 6.0) soils must be limed to raise their pH to as close to neutral (7.0) as possible; however, this may be challenging in highly acidic soils unless very large lime inputs are used, along with plenty of time.

Copper (Cu)

Due to the fact that fungicides containing copper are one of the most effective tools in the fight against Downy mildew, copper is a trace element that vines need in very small amounts and is rarely in short supply. In areas where relatively antiquated copper sprays, like Bordeaux Mixture, have been used frequently, situations of excess copper have been documented, and vines may become stunted. When copper is utilized excessively, earthworm and microbial activity in the soil are also significantly decreased. Often, the application of lime and humus, along with green manuring and alternative fungicides, will address an overabundance of copper.

Iron (Fe)

As previously noted, iron is necessary for the synthesis of chlorophyll, and iron deficiency is typically only found in extremely alkaline (high pH) soils. As a temporary fix, liquid fertilizers containing chelated iron—iron that has been processed to make it soluble—can be sprayed on vines or injected into the ground. If vines growing on highly alkaline soils are to avoid lime-induced chlorosis, the right rootstock selection is imperative in the long run.

Magnesium (Mg)

A component of the chlorophyll molecule, magnesium is a minor element but a crucial one in the process of photosynthesis. Magnesium-deficient vines will exhibit chlorosis symptoms, such as yellowing leaves that become more apparent as the vine ripens. Deficits in magnesium are also linked to bunch stem necrosis, which damages bunch stems and hinders ripening (see later in this chapter). Even though there may be sufficient magnesium in the soil, symptoms of a magnesium deficiency are often seen in potassium-rich soils. Certain rootstocks, like Fercal and SO4, don't seem to be very good at absorbing magnesium, and vines that grow on them—which are frequently planted in chalky, high-pH soils—will exhibit symptoms of magnesium deficiency. Short-term foliar feeds (Epsom salts work wonders) and long-term fertilizers are among the remedies. Magnesium deficiencies are typically greater in light, sandy soils than in heavier clay or loam soils.

Manganese (Mn)

Manganese is a trace element that the vine needs in very small amounts in order to grow successfully. Deficiencies are most common in alkaline, sandy soils and are indicated by yellow stripes between the leaf veins. Giving the vine timely foliar feeds is typically successful in giving it enough manganese. Excess manganese can occasionally be an issue in acidic soils, resulting in crop loss and stunted growth. Chlorosis is another common issue that befalls vines in these soils, and the two conditions frequently coexist. Molybdenum (Mo) is a metal that is present in trace amounts in plants, particularly vines. Because applying molybdenum doesn't cost much and has been linked to better flowering (particularly in Australia with Merlot), pre-flowering sprays frequently include applications of "moly." Molybdenum levels should not be raised too high, and soils should be tested frequently.

Sulfur (S)

Sulfur is an essential element for all plants, but fortunately, vines rarely experience deficiencies because super-phosphate fertilizers, which contain 11% sulfur, and sulfur sprays against powdery mildew (Oidium) are used. Zinc (Zn) Soils with high phosphorus levels or sandy, alkaline soils tend to have zinc deficiencies. The first signs of a zinc deficit are yellow veining and asymmetrical leaf growth patterns. Stunted shoot growth and inadequate fruit set follow. Zinc-deficient vines have reduced capacity to produce auxins, plant hormones that stimulate cell division. Zinc can be sprayed on leaves and, in vineyards that are vulnerable, it can be applied frequently during the growing season.

BSN Bunch Stem Necrosis: Numerous factors have been linked to bunch stem necrosis, including overly vigorous growth, inadequate nutrition (low calcium and magnesium levels), and cold and wet weather during flowering. Low sugar and high acid results from the condition, which also causes bunches to shrivel and the berries to stop developing. Also, yields are drastically decreased. When called Early Bunch Stem Necrosis, or EBSN, the issue can also impact vines prior to flowering. Beyond addressing the deficiencies in nutrition, there isn't a true treatment for BSN.

Iron-Chlorosis, Lime-Induced Chlorosis

When vines are planted in high-lime soils, iron—which the vine needs to produce chlorophyll, which gives leaves their green color and is essential for photosynthesis—becomes trapped in the soil and is unable to be absorbed by the plant. Testing is necessary to determine the percentage of active calcium carbonate (calcaire actif), which is different from the total calcium carbonate content in high-lime or chalk-containing soils (typically those with pH values of 7.5 or higher). This can vary from 0% to 50%, and if it exceeds 5%, a suitable rootstock that can withstand lime should be chosen.

Coulure

Poor flowering conditions and/or an imbalance in the nutrients in the vine can cause coulure. Bunches with few berries result from poor or imperfect pollination caused by one or both of these factors. This typically translates into a significant crop loss. While some clones are less prone to coulure, certain varieties are, and methods like tip trimming during flowering may be helpful. This condition is also occasionally referred to as "shatter" in English.

NOTE

Printed in Great Britain
by Amazon

797f10d8-055b-4308-9657-0712ed848297R01